46 Advances in Biochemical Engineering/ Biotechnology

Managing Editor: A. Fiechter

Modern Biochemical Engineering

Guest Editor: T. Scheper

Dedicated to Prof. Karl Schügerl
on the occasion of his 65th birthday

With contribution by
D. J. Barnes, I. Economidis, I. Endo, A. Fiechter,
S. Furusaki, G. E. Grampp, H. Håkanson,
M. N. Karim, I. Karube, B. Mattiasson,
J. Nielsen, S. L. Rivera, A. Sambanis, M. Seki,
S. Shioya, B. Sonnleitner, G. N. Stephanopoulos,
T. Takeuchi

With 96 Figures and 27 Tables

Springer-Verlag
Berlin Heidelberg GmbH

ISBN 978-3-662-14991-1   ISBN 978-3-540-47005-2 (eBook)
DOI 10.1007/978-3-540-47005-2

© Springer-Verlag Berlin Heidelberg 1992
Originally published by Springer-Verlag Berlin Heidelberg New York in 1992
Softcover reprint of the hardcover 1st edition 1992

Library of Congress Catalog Coard Number 72-152360

Typesetting: Th. Müntzer, Bad Langensalza;

02/3020-5 4 3 2 1 0 — Printed on acid-free paper

Managing Editor

Professor Dr. A. Fiechter
Institut für Biotechnologie, Eidgenössische Technische Hochschule
ETH — Hönggerberg, CH-8093 Zürich

Guest Editor

Dr. T. Scheper
Institut für Technische Chemie, Universität Hannover
Callinstr. 3, D-3000 Hannover

Editorial Board

Prof. Dr. Karl Schügerl

To
Professor Dr. Karl Schügerl on his 65th Birthday

On 22nd June 1992, Karl Schügerl will spend his 65th birthday in the midst of a large assembly of students, PhD students and colleagues in the Faculty of Science. For him, as well as for the Institute of Chemical Engineering of the University of Hanover, it is a day of honour. Those involved in chemical and biological process engineering will join in our congratulations to him on the day, and in our best wishes to him for a happy future.

Karl Schügerl has built up an exemplary career, which began after he completed his chemical engineering studies at Budapest University in 1949. Since that time he has continued to devote himself to research, either in industry or at university. The first steps in his career were in organic chemistry at Budapest University (1949–52); experimental institutions in the organic-chemical industry, Budapest (1952–55); a design studio for bioplant (1955–56) and at Riedel de Haen AG in Seelze, Germany (1956–58). He later completed his PhD degree at Hanover University in the kinetics and rheology of fluid bed systems and spent three years of post-doctorate work at the Universities of New York (high-temperature pyrolysis of hydrocarbons) and Princeton in the USA, where he was closely involved for one and a half years in fundamental investigations of molecular beams for aerospace research. Soon after his return to Germany, he qualified for inauguration as a lecturer in Hanover and shortly afterwards accepted a professorship in process engineering at Braunschweig University. In 1969, he finally returned to Hanover University. In due course, the purposeful nature of his work developed the Institute for Chemical Engineering into a research institute concentrating on biological process engineering. His early work on high-temperature pyrolysis of olefins and on molecular beam ultrasonic applications opened up typical problem areas relating to mass transfer in chemical processes. Hydrodynamics in single drops, three-phase *fluid beds, cryosorption, residence time distributions, test material mixing, and flash photolysis were typical fields of work in the 1960s and 1970s. Hundreds of experimental publications bear witness to his comprehensive knowledge in these fields. The urge towards constantly new fields of work led Karl Schügerl in 1975 into biotechnology, in which today his institute occupies a dominant position. It has taken a leading role particularly in on-line measurement and control of biological process engineering. Karl Schügerl was one of the first chemical engineers in Germany to conduct extensive analyses of biological aspects. These also include hundreds of papers in the biological field which omit hardly any of the major measuring techniques or process applications. Bacteria, yeasts or fungi are the main agents in this highly developed discipline. New solutions to the problems of future improvements in performance are sought using the most modern methods of analysis and software. His present areas of interest have recently even extended beyond biological process engineering and today cover environmental technology, biosensors, the disposal of highly polluted effluent, soil renewal and hydrometallurgy.*

Karl Schügerl has always refused illustrious offers from renowned universities, thereby dedicating himself to developments in Hanover; this was certainly the result of his important recognition that only determined efforts would lead to progress in the rapidly developing field of biotechnology.

It would seem significant in this connection that he willingly accepts many calls on his time to act as consultant and expert. He acts as adviser to countless institutions both in Germany and abroad in the academic and public domain, where his reliable advice and support are highly esteemed. Between 1982 and 1986, he administered the department for biological process engineering at the GBF (Association for Biotechnological Research) in Braunschweig.

Today, 700 publications, including five books, bear witness to the iron discipline by which Karl Schügerl has developed his view of things. Nevertheless he has remained an approachable and cherished colleague, who always finds time for scientific requests. This also holds true for the next generation of scientists for whom he not only proposes attractive subject areas, but also provides support as a helpful adviser in professional and personal problems. There are many of his former PhD students who today are themselves in academic positions, and who are able to pass on the enthusiasm for science of their former teacher to the coming generations. In industry too, graduates from Hanover have become successful and, with their knowledge and skill, are able to support their companies in these times of stiff economic competition.

Recently, the academic and scientific services of our honoured colleague have been recognized by Budapest University with an honorary doctorate.

This present publication allows the authors to add their appreciation. Their contributions demonstrate the wide interests of Karl Schügerl, which cover many subsidiary fields. We hope in this way to give him great pleasure and thereby to demonstrate the high esteem in which we hold him.

Those who are familiar with the great creative powers of this gifted scientist cannot imagine any slackening of his efforts in the promotion of research and education. We offer our good wishes for his health and the best conditions for the accomplishment of his plans.

Armin Fiechter

Attention all "Enzyme Handbook" Users:

A file with the complete volume indexes Vols. 1 through 5 in delimited ASCII format is available for downloading at no charge from the Springer EARN mailbox. Delimited ASCII format can be imported into most databanks.

The file has been compressed using the popular shareware program "PKZIP" (Trademark of PKware INc., PKZIP is available from most BBS and shareware distributors).

This file is distributed without any expressed or implied warranty.

To receive this file send an e-mail message to: SVSERV@DHDSPRI6.BITNET.
The message must be: "GET/ENZHB/ENZ_HB.ZIP".

SPSERV is an automatic data distribution system. It responds to your message. The following commands are available:

HELP	returns a detailed instruction set for the use of SVSERV,
DIR *(name)*	returns a list of files available in the directory "name",
INDEX *(name)*	same as "DIR"
CD ⟨*name*⟩	changes to directory "name",
SEND ⟨*filename*⟩	invokes a message with the file "filename",
GET ⟨*filename*⟩	same as "SEND".

Table of Contents

Artificial Neural Networks in Bioprocess State Estimation

M. N. Karim and S. L. Rivera
Department of Agricultural and Chemical Engineering, Colorado State University,
Fort Collins, CO. 80523, U.S.A.

The application of artificial neural networks to the estimation and prediction of bioprocess variables is presented in this paper. A neural network methodology is discussed, which uses environmental and physiological information available from on-line sensors, to estimate concentration of species in the bioreactor. Two case studies are presented, both based on the ethanol production by *Zymomonas mobilis*. An efficient optimization algorithm which reduces the number of iterations required for convergence is proposed. Results are presented for different training sets and different training methodologies. It is shown that the neural network estimator provides good on-line bioprocess state estimations.

Advances in Biochemical Engineering
Biotechnology, Vol. 46
Managing Editor: A. Fiechter
© Springer-Verlag Berlin Heidelberg 1992

List of Symbols

In Z. mobilis Kinetic Model:

a		Power of the ethanol inhibition term in μ
b		Power of the ethanol inhibition term in q_p
D	h^{-1}	Dilution rate
F	$l\,h^{-1}$	Permeate flow
F_0	$l\,h^{-1}$	Total feed flow to the system
K_i	$g\,l^{-1}$	Substrate inhibition constant for growth
K_i'	$g\,l^{-1}$	Substrate inhibition constant for ethanol production
K_s	$g\,l^{-1}$	Monod kinetic constant
K_s'	$g\,l^{-1}$	Saturation constant for q_p
p	$g\,l^{-1}$	Ethanol concentration
P_i'	$g\,l^{-1}$	Ethanol threshold concentration for ethanol production
P_m	$g\,l^{-1}$	Maximum ethanol concentration for cell growth
P_m'	$g\,l^{-1}$	Maximum ethanol concentration for ethanol production
q_p	$g\,g^{-1}\,h^{-1}$	Specific ethanol production rate
q_{pm}	$g\,g^{-1}\,h^{-1}$	Maximum specific ethanol production rate
R		Recycle ratio
s	$g\,l^{-1}$	Glucose concentration
S_i	$g\,l^{-1}$	Threshold substrate concentration for cell growth
s_o	$g\,l^{-1}$	Glucose concentration in the feed stream
x	$g\,l^{-1}$	Biomass concentration
$x_{\max}$	$g\,l^{-1}$	Maximum cell concentration
$Y_{p/s}$	$g\,g^{-1}$	Ethanol yield, g ethanol per g substrate consumed
μ	h^{-1}	Specific growth rate
μ_{mo}	h^{-1}	Maximum specific growth rate at zero ethanol concentrations

In Neural Network Theory:

$\mathbf{d}_p$	L-dimensional target vector
E_p	Sum-of-squares error for training example p
$E(\mathbf{w})$	Total sum-of-squares error for all input patterns, function of weight vector $\mathbf{w}$
$\mathbf{G}^{(q)}$	Vector of steepest descent directions in iteration (q)
$\mathbf{g}_p^{(q)}$	Gradient vector of one input-output pattern p
$g_{uv}^{(q)}$	Element of the gradient vector $\mathbf{g}_p^{(q)}$
n	Number of interconnection weights in the network
O_{pj}	Output of neuron j from the training set p
p	number of training pattern
q	number of iteration
$\mathbf{r}^{(q)}$	Vector of conjugate gradient directions in Eq. (17)
S_{pj}	Activation state of neuron j from the training set p
$\mathbf{S}^{(q)}$	Vector of search directions for conjugate gradient algorithm
$\mathbf{w}^{(q)}$	Vector of neural network weights in iteration q

w_{ij}	Interconnection weight from node i to node j
$\mathbf{x}_p$	N-dimensional network input vector
$\mathbf{y}_p$	L-dimensional network output vector
$\mathbf{z}^{(q)}$	Vector of conjugate gradient directions in Eq. (16)
$\alpha^{(q)}$	Step size in iteration q used in Eq. (8)
$\beta^{(q)}$	Step size in iteration q used in Eq. (17)
$\gamma^{(q)}$	Step size in iteration q used in Eq. (16)
δ_{pv}	Change in E_p due to changes in neural activation state S_{pu}.
σ	Argument of the sigmoid function.

1 Introduction

The successful operation, control and optimization of bioprocesses rely heavily on the availability of a fast and accurate evaluation of the system performance. This in turn requires reliable real-time process variable information. Direct on-line measurements of primary process variables, such as biomass, substrate and product concentrations, usually are unavailable. Most industrial bioprocess control policies are based upon the use of infrequent off-line sample analysis, leading to poor process operability and regulation. The state of the cultivation, therefore, has to be inferred from measurements of secondary variables and any previous knowledge of the process dynamics.

In recent years, several techniques have been developed for indirect estimation of process variables and data analysis of biological systems. Approaches range from statistical methods, including linear and nonlinear regression, to artificial intelligence techniques, for determination of rules for expert systems from on-line data. Estimation techniques generally are used to predict unmeasurable process variables, and/or to identify process model parameters, by establishing the structure of a process model. However, it is well-known that models of real nonlinear systems possess a great amount of uncertainty. This is due in part to imprecise measuring devices, environmentally dependent system parameters, or disturbances inherent in the plant. However, the most important cause of process/model mismatch is perhaps our incomplete knowledge of the system dynamics. The quality of the estimation depends greatly on the depth of understanding of the process. Methods which can provide an adequate estimation of process states and parameters in spite of incomplete process knowledge could be successfully applied in control and optimization of bioprocesses.

The use of artificial neural nets for identification and control of chemical and biochemical plants has recently been the focus of some research groups [1, 2, 3, 4, 5]. Artificial neural nets are highly interconnected networks of non-linear processing units arranged in layers and having adjustable connection strengths (weights) [6]. Neural networks can approximate large classes of nonlinear functions by changing the strengths of the connections on the links, a procedure which is called *learning*. Neural networks are well-known for their ability to adjust dynamically to environmental changes, to perform useful generalization from specific examples, and to find relevant regularities in the data. Recent applications of neural nets have been mostly in the areas of speech recognition, image processing, optimization problems, robotics, decision making, and identification and control. However, no non-trivial experimental validation of state estimation in biological systems using neural networks has been reported in the literature.

In this paper, we will present some applications of artificial neural networks to the estimation and prediction of bioprocess variables. Biological kinetic models seldom include the influence of environmental factors, like pH and temperature. The development of quantitative models which simulate how cells respond to various environmental changes, will help in better utilizing the chemical synthesis capabilities of the cells. We will discuss a neural network methodology which will use environmental and physiological information available from on-line sensors, to estimate concentrations of species in the bioreactor.

Two case studies will be presented, both based on ethanol production by *Zymomonas mobilis*. Yeasts and bacteria have been successfully used in the past for bulk production of ethanol. Reported data on ethanol productivities for yeasts and *Z. mobilis* have demonstrated that *Z. mobilis* is superior to yeasts as far as productivity is concerned [7]. Both the specific ethanol productivity and specific glucose uptake rate are several times larger in *Z. mobilis* than in yeasts; this is associated with lower levels of biomass formation during the cultivation. Given the obvious advantages of *Z. mobilis* for industrial ethanol production, it is worthwhile to investigate ways of improving the bioprocess evaluation, by means of better process variable predictions. Neural networks applied in this context will demonstrate to be a powerful tool for identification and control in biological systems.

2 Estimation Methods in Bioprocesses

In terms of control and optimization, bioprocesses are not more complex or unique than any other chemical process. However, species are ill-defined and interact, and the kinetics is poorly understood [8]. Both process and disturbance dynamics are uncertain and variable. This lack of knowledge of the underlying principles, together with the scarcity of appropriate biosensors, is what makes the identification procedure so difficult.

Different estimation-filtering techniques have been developed to aid in the prediction of the biological and physicochemical parameters needed for control and regulation. Identification techniques may be divided into *parameter estimation* and *functional estimation* [9]. The parameter estimation approach is simple, if one can safely assume a known system function with only unknown parameters. Functional estimation deals with the estimation of the system function, as well as its parameters.

Model-based estimation methods, are parameter estimation techniques, where usually simplified nonlinear kinetic models and/or material balance equations [10,

11] are defined. This approach is useful in cases when all the species involved are known and their elemental composition is completely defined and time-invariant. Otherwise, changes in growth rates or in the nature and composition of the medium may affect the chemical composition of the cells [12], invalidating the process description.

Other parameter estimation techniques include the least-squares fitting of linear, time series models, valid only in a certain operating range. The success of this approach depends on the match between the actual system dynamics and the linearized approximation. Adaptation can provide some improvement in the estimation, but in general, these models do not represent the dynamic nonlinear behavior of the process accurately. Several examples of this approach exist in the literature [13, 14, 15].

Stephanopoulos and San [16] integrated the concept of material balance with estimation using a Kalman filter, for the simultaneous state and parameter estimation of a bioprocess system. Kalman filtering is a technique for estimating the underlying trend in noisy measured data. Due to the iterative nature of the filter, an inaccesible state variable can be estimated via an imperfect process model [8]. An extension to the Kalman filtering technique involves the use of nonlinear models and their subsequent linearization. Several applications of the Extended Kalman Filter (EKF) for state estimation in biotechnological processes have been reported [17, 18, 19]. However, in spite of the apparent success of this widely used estimation technique, care should be exercised due to its inherent limitations. Assumptions of the EKF include the use of a linear process model (or a linearized version of a nonlinear model), previous knowledge of the covariance matrices of the state and output noise, and a good initial value for the filter covariance. Sensitivity problems in biological systems due to deviations from this assumptions may cause the performance of the estimator to deteriorate, resulting in biased estimates of the states.

Since the biochemical process is non-linear, a better estimation is to be expected from exploiting a nonlinear structure for designing the estimator. Bastin and Dochain [20] applied asymptotic observers for the estimation of specific growth rate. These are a class of estimators that can handle nonlinearities in the process model. Improvements to this algorithm have been proposed [21], but two main disadvantages remain: high sensitivity to wrong initial guess of the biomass concentration and the use of off-line measurements to define the state of the process.

Therefore, it appears that a major improvement in bioprocess control and optimization can be made if one has an estimation algorithm which can effectively reduce the off-line analysis frequency, while maintaining the quality of the information available.

3 Overview of Neural Networks

Neural networks offer the opportunity to directly model nonlinear processes, and estimate or predict the values of relevant process variables. Artificial Neural Networks (ANNs) models consist of a large number of simple interconnected nonlinear computational elements *(neurons)*. The strength of the links between nodes is typically adapted during use to improve performance. Neural networks have received wide interest during the last years due to their abilities of learning, generalization and abstraction.

At present, in chemical and biochemical engineering applications, the most widely used neural net training method is *backpropagation*, a generalization of the Least Mean Square Error algorithm. Backpropagation uses an iterative gradient search technique to minimize a performance function, equal to the mean square difference between the desired and the actual net outputs. Several applications of ANNs trained by backpropagation have been reported in the literature.

ANNs have been applied as a method for identifying regular patterns in data. Qian et al. [22] presented a new method for predicting the secondary structure of globular proteins based on nonlinear neural network models. Network models learned from existing protein structures how to predict the secondary structure of local sequences of amino acids. The use of this technique provided a significant improvement over existing methods for nonhomologous proteins. In another application, McAvoy et al. [23] used neural nets to deconvolute fluorescence spectra obtained from solutions of amino acid fluorophores. They showed that neural nets are superior to standard linear techniques in estimating the solution concentrations, due to the adequate handling of the nonlinearities in the spectral data.

Neural networks are well-suited to learn and retrieve correlations between measurements and faults or responses, and are particularly useful when the measurements are incomplete or inaccurate [2, 24]. A technique for the application of neural nets for diagnosing process failures during process transients has been recently investigated [3]. The symptom patterns were presented to the network in two ways: by using a raw time-series process sensor data and by using a moving average value of the same time-series data. Both methods were successful in detecting simple and interactive faults, even when they were trained only in single fault cases. This demonstrates the neural networks ability for generalization.

Other recent research deals with the feasibility of using neural networks for estimation and control of nonlinear plants. Bhat et al. [5] investigated the use of neural nets trained by backpropagation to estimate and control the variables of a simulated pH CSTR. They extended the backpropagation approach to enable the dynamic modeling of a nonlinear process by including past values of process variables as inputs to the network. Their results showed a better overall estimation compared to traditional linear ARMA (autoregressive moving average) modeling. This approach, using a *temporal* or *sliding window*, has also been investigated by other researchers [25, 26, 27]. Since only simulation analysis were performed, it was assumed that all the required process variables were available on-line, an assumption which is not always valid. One of the disadvantages of using *temporal windows* is that process delays should be known, so that an adequate number of past observations can be used, in order to include the influence of all input variations on the prediction of the outputs.

Adaptive learning is one of the most attractive features of neural networks, since they learn how to perform certain tasks by undergoing training with illustrative examples. However, it is important to realize that when building an empirical model from plant data, one only obtains a model for the correlation structure that was present when the data was collected. The problem lies with the data set. Real process data usually contain many variables that are highly correlated. To overcome this problem, and be able to use the correlation models to imply cause and effect relationships one must design the experiments, or use causal mechanistic models (MacGregor et al. [28]). A training data set should be carefully selected such that it will provide sufficient information of the behavior of the system.

4 Theory of Neural Networks

The objective of training the network is to adjust the interconnection weights so that the application of a set of inputs produces the desired set of outputs (Fig. 1). Inputs to the neural network estimator may consist of the manipulative inputs to the plant (temperature, pH, flow rates, etc.), together with other measured process variables. These measurements, such as redox potential and carbon dioxide evolution rate, provide information about the state of the process. The corresponding process outputs (biomass, ethanol and glucose concentrations) provide the desired *teacher* signal, which trains the network. The difference between the desired output and the value predicted by the network is the prediction error. Iterations are performed to minimize the total prediction error by modifying the values of the interconnection weights until convergence is achieved.

Consider the layered feedforward neural network in Fig. 2. It consists of one input layer, one hidden layer and one output layer. In general, it may consist of more than one hidden layer. An N-dimensional input vector $\mathbf{x}_p$ is presented to the network and an L-dimensional vector $\mathbf{y}_p$ is obtained from the output layer.

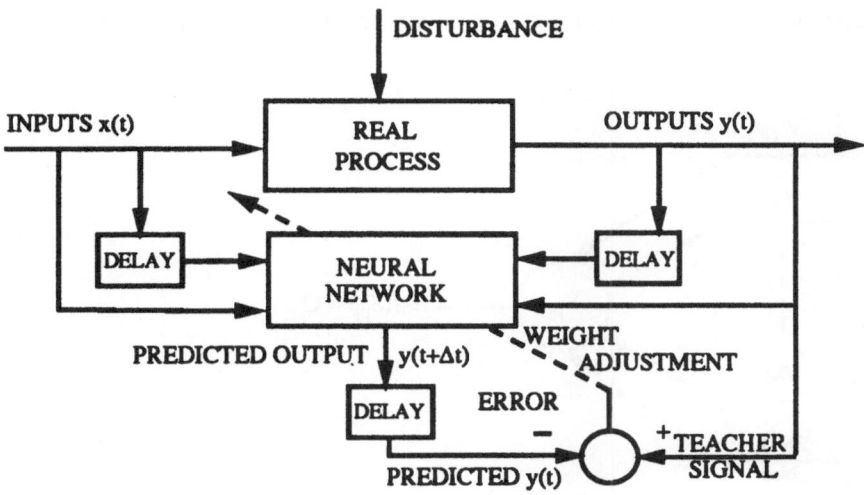

Fig. 1. Block diagram of the neural network estimator

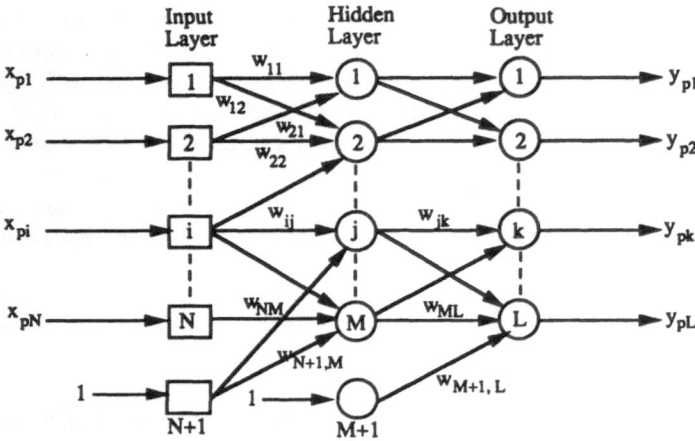

Fig. 2. Two-layer neural network (not all interconnections are shown)

All inputs and outputs to the network have to be scaled in the range 0 to 1, so that all inputs have equal weight in the calculations regardless of their order of magnitude. The nodes in the input layer (represented by boxes) perform this preconditioning of the input vector. The circles in the hidden and output layers are the neurons or computational elements, and the lines represent the interconnection weights. With the exception of the output layer, the layers should contain a bias node whose value is 1. This fixed unit allows each neuron to find its own threshold value through the training procedure.

A typical neuron performs two functions: a weighted linear combination of its input component (activity) and a nonlinear transformation of this activity value. Figure 3 shows the j-th neuron in the hidden layer. The inputs to this neuron consist of the N-dimensional vector $\mathbf{x}_p$ and a unit bias. Each input is multiplied

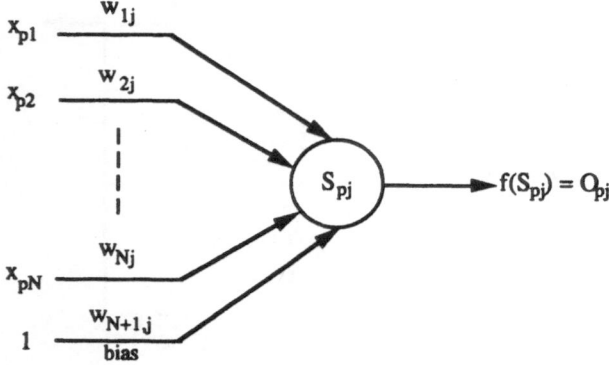

Fig. 3. An artificial neuron

by a weight w_{ij} and the products are summed up to give the activation state S_{pj}:

$$S_{pj} = \sum_{i=1}^{N} w_{ij}x_{pi} + w_{N+1,j} \tag{1}$$

The output of the neuron O_{pj} is then calculated as the sigmoid function of the activation state:

$$O_{pj} = f(S_{pj}) = \frac{1}{1 + e^{-S_{pj}}} \tag{2}$$

where f represents the differentiable and nondecreasing sigmoid function (Fig. 4) of σ, with derivative

$$f'(\sigma) = f(\sigma)\,[1 - f(\sigma)] \tag{3}$$

The output layer performs the same calculations given by Eqs. (1) and (2), except that the input vector $\mathbf{x}_p$ is replaced by the hidden layer output vector $\mathbf{O}_p$ and the weights required are w_{jk}:

$$S_{pk} = \sum_{j=1}^{M} w_{jk}O_{pj} + w_{M+1,k} \tag{4}$$

$$O_{pk} = y_{pk} = f(S_{pk}) = \frac{1}{1 + e^{-S_{pk}}} \tag{5}$$

The set of training examples consists of P input/output vector pairs $(\mathbf{x}_p, \mathbf{d}_p)$. Weights are initially randomized. Thereafter, weights are selected so as to minimize

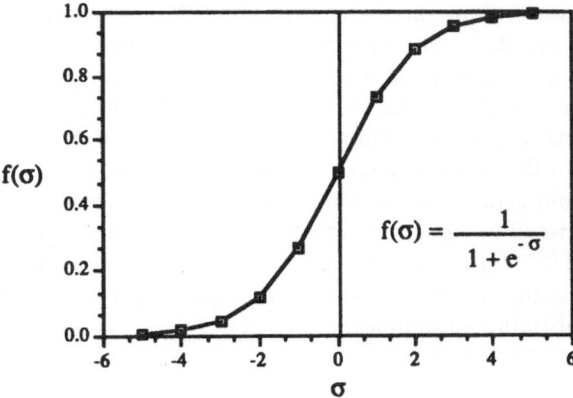

Fig. 4. Sigmoid function

the objective function $E(\mathbf{w})$, defined as the mean squared error between the actual outputs from the output layer y_{pk} and the desired or target outputs d_{pk} (measured outputs from the plant) for all the input patterns:

$$E(\mathbf{w}) = \sum_{p=1}^{P} E_p \qquad (6)$$

where E_p is the sum-of-squares error associated with a training example:

$$E_p = \tfrac{1}{2} \sum_{k=1}^{L} (d_{pk} - y_{pk})^2 \qquad (7)$$

The network weights are arranged in a vector format, and then modified to minimize the objective function $E(\mathbf{w})$. A nonlinear numerical optimization can be used to minimize this nonlinear multivariable objective function. The backpropagation algorithm, which is the traditional training algorithm, optimizes the objective function by the gradient method of steepest descent [6]. However, as it has been shown in other applications [29], steepest descent is highly inefficient as an optimization method, requiring many iterations for convergence and lacking robustness. Instead, the training algorithm in this study will use Le's Conjugate Gradient (CG-LE) [30] method for unconstrained optimization. This method has shown better convergence, stability and speed than other conjugate gradient or quasi-Newton methods of optimization, and at the same time, requires less computational storage.

The CG-LE method is a type of conjugate gradient method with dynamic optimization of step size $\alpha^{(q)}$. In each iteration q, the vector of interconnection weights $\mathbf{w}$ is improved by:

$$\mathbf{w}^{(q+1)} = \mathbf{w}^{(q)} + \alpha^{(q)} \mathbf{S}^{(q)} \qquad (8)$$

where $\alpha^{(q)}$ is chosen to minimize $\mathbf{w}$ along the search direction $\mathbf{S}^{(q)}$. The algorithm generates n mutually conjugate directions and minimizes a positive definite quadratic function of dimension n in at most n steps, where n is the size of the vector $\mathbf{w}$. An important feature, however, is that it is independent of any quadratic assumptions when minimizing a general nonlinear function. Convergence is assumed when each element of the gradient vector is less than or equal to 10^{-6}.

In common with other conjugate gradient methods, a sequence of search directions $\mathbf{S}^{(q)}$ is generated, formed by linear combinations of $-\mathbf{G}^{(q)}$ (the current steepest descent direction) and $\mathbf{S}^{(q-1)}$ (the previous search direction). The direction of steepest descent with respect to the overall objective E, is given by the sum of the gradients of the individual training examples:

$$\mathbf{G}^{(q)} = \sum_{p=1}^{P} \mathbf{g}_p^{(q)} \qquad (9)$$

The gradient $\mathbf{g}_p^{(q)}$ is calculated with respect to one input-output pair p, its components $\mathbf{g}_{uv}$ being:

$$\mathbf{g}_{uv}^{(q)} = \frac{\partial E_p}{\partial w_{uv}^{(q)}} \tag{10}$$

representing the direction of steepest descent of the weight w_{uv} from node u to node v with respect to the partial objective E_p. By applying the chain rule. Eq. (10) can be written as:

$$\frac{\partial E_p}{\partial w_{uv}^{(q)}} = \frac{\partial E_p}{\partial S_{pu}} \frac{\partial S_{pu}}{\partial w_{uv}} = -\delta_{pv} O_{pu} \tag{11}$$

Therefore, the delta values δ_{pv} represent the change in E_p due to changes in neural activation state S_{pu}, and O_{pu} is the change in neural activation state S_{pu} due to changes in w_{uv}.

Backpropagation provides two rules for calculating the error signal of a node, depending on whether the neuron is in the output layer or in a hidden layer. For calculating δ_{pk} between the hidden layer neurons to the output neuron k:

$$\delta_{pk} = (d_{pk} - y_{pk}) f'(S_{pk}) \tag{12}$$

where f' is defined according to Eq. (3). In a similar manner, δ_{pj} can be calculated for the input layer neurons to the hidden neuron j.

$$\delta_{pj} = f'(S_{pj}) \sum_{k=1}^{L} \delta_{pk} w_{jk} \tag{13}$$

Hoskins et al. [24] summarizes the learning procedure as consisting of two phases. First, the inputs are propagated through the network in a feedforward fashion to produce output values that are compared to the desired outputs, resulting in an error signal for each of the output nodes. Second, the error is propagated backward through the network. The delta values are calculated first for the output layer, and these delta values are used recursively to calculate the deltas for the adjacent hidden layer using Eqs. (12) and (13). These deltas can be used to update all the weights in the network.

The CG-LE optimization algorithm is described as follows. Given the initial weight vector $\mathbf{w}^{(0)}$, let $\mathbf{S}^{(0)} = -\mathbf{G}^{(0)}/\|-\mathbf{G}^{(0)}\|$, the vector norms being Euclidean. For $q \geq 0$, the optimization step during the iteration q is given by:

$$\mathbf{w}^{(q+1)} = \mathbf{w}^{(q)} + \alpha^{(q)} \mathbf{S}^{(q)} \tag{14}$$

where $\alpha^{(q)}$ is the value of α minimizing the overall network error function $E(\mathbf{w}^{(q)} + \alpha \mathbf{S}^{(q)})$. If $\mathbf{G}^{(q+1)} = 0$, stop; otherwise, let

$$\mathbf{S}^{(q+1)} = \frac{(\mathbf{z}^{(q+1)} - \mathbf{w}^{(q+1)})}{\|\mathbf{z}^{(q+1)} - \mathbf{w}^{(q+1)}\|} \tag{15}$$

where:

$$z^{(q+1)} = r^{(q+1)} + \gamma^{(q+1)}S^{(q)} \tag{16}$$

$$r^{(q+1)} = w^{(q+1)} + \beta^{(q+1)}\left(\frac{-G^{(q+1)}}{\|G^{(q+1)}\|}\right) \tag{17}$$

where $\gamma^{(q+1)}$ minimizes $E(r^{(q+1)} + \gamma S^{(q)})$ and $\beta^{(q+1)}$ is chosen according to an inexact line search method proposed by Le [30].

As in other conjugate gradient methods, a restart criterion has to be proposed, since the search direction's convergence becomes linearly dependent after some iterations. To improve the rate of convergence on a general nonlinear function, the algorithm is restarted occasionally by using the direction of steepest descent $S^{(q)} = -G^{(q)}$ instead of Eq. (15). The criterion is simply to restart the algorithm with the steepest descent direction if the inequality

$$\|w^{(q+1)} - w^{(q)}\| \leqq \|z^{(q)} - w^{(q)}\| \tag{18}$$

is satisfied for two consecutive iterations, and the total number of iterations counted from the previous restart is greater than or equal to $n + 1$.

4.1 The Training Algorithm

Step 0. Initialize iteration number $q = 0$ and training set number $p = 0$.

Step 1. Set all weights in the network to small random numbers.

Step 2. Select the next training pair (x_p, y_p) from the training set $(p = p + 1)$; apply the input vector to the network input layer. Set the scale for normalizing the inputs between 0 and 1.

Step 3. Calculate actual outputs of the hidden layer and output layer. Use Eqs. (1), (2), (4) and (5).

Step 4. Obtain the error gradient g_p for the training pair p:

(a) Calculate the error between the network output and the desired output (target vector from the training pair p) with $e_{pk} = d_{pk} - y_{pk}$ for $k = 1, ..., L$.

(b) Calculate the L delta values for the hidden layer nodes to output neurons using Eq. (12).

(c) Calculate the M delta values for the input layer nodes to the hidden layer neurons using Eq. (13). Use the w_{jk} weights from the previous iteration $(q - 1)$.

(d) Calculate the gradient $g_{jk} = -\delta_{pk}O_{pj}$ according to Eq. (11), for $j = 1, ..., M + 1$ and $k = 1, ..., M$, and where $O_{p,M+1} = 1.0$.

(e) Calculate the gradient $g_{ij} = -\delta_{pj}O_{pi}$ according to Eq. (11), for $i = 1, ..., N + 1$ and $j = 1, ..., M$, where P_{pi} is the scaled input value x_{pi} and $O_{p,N+1} = 1.0$.

(f) Arrange all gradients in a vector of gradients $\mathbf{g}_p^{(q)}$ and save.

(g) If $p = P$ (all training pairs have been processed), continue otherwise go to Step 4.

Step 5. Calculate the gradient $\mathbf{G}^{(q)}$ with respect to the overall objective function E by using Eq. (9).

Step 6. Calculate the Euclidean norm of the total gradient vector $\mathbf{G}^{(q)}$ as follows:

$$\| \mathbf{G}^{(q)} \| = \left(\sum_i \mathbf{G}_i^2 \right)^{1/2} \tag{19}$$

Step 7. Check for convergence. Each element of the gradient vector $\mathbf{G}^{(q)}$ should be less than or equal to a tolerance value. If true, then network is trained. Otherwise, continue to Step 8.

Step 8. Determine search direction according to CG-LE unconstrained optimization methodology:

(a) If $q = 0$, calculate the search direction $\mathbf{S}^{(0)} = -\mathbf{G}^{(0)}/\|\mathbf{G}^{(0)}\|$.

(b) If $q \neq 0$, then
- Perform coarse line search to rescale initial stepsize $\beta^{(q+1)}$.
- Calculate $\mathbf{r}^{(q+1)}$ according to Eq. (17).
- Perform line search to find $\gamma^{(q+1)}$ minimizing the error function $E(\mathbf{r}^{(q+1)} + \gamma \mathbf{S}^{(q)})$.
- Calculate $\mathbf{z}^{(q+1)}$ by Eq. (16).
- Calculate the search direction $\mathbf{S}^{(q+1)}$ according to Eq. (15).

Step 9. Perform line search procedure to obtain α minimizing the error function $E(\mathbf{w}^{(q)} + \alpha \mathbf{S}^{(q)})$.

Step 10. Calculate new weight vector $\mathbf{w}^{(q+1)}$ from Eq. (14). Update iteration number $q = q + 1$.

Step 11. Propagate new weights through the network. Goto Step 2.

4.2. Outline of the Estimation Scheme

The detailed procedure for developing and executing a neural network estimator is as follows:

1. Determine available measurements (on-line or infrequent off-line) to be used by the estimator. This will define the number of inputs N.
2. Determine which process variables should be estimated and/or predicted. This will define the number of outputs L.
3. Define a training set consisting of P pairs of input/output vectors. If possible, subject input to pseudo-random binary sequence and generate output at discrete output intervals.
4. Select a network configuration for N inputs and L outputs. The use of one hidden layer is recommended. Specify number of hidden nodes.
5. Train the network off-line using the training set.
6. Test the network by presenting a set of input vectors and observing predicted output vectors. Preferably this test set should not belong to the training set.
7. Calculate error difference between predicted outputs and real outputs. Obtain mean square error (SSE) as a performance index for the network.

8. If mean squared error is acceptable, continue. Otherwise, use a different configuration by repeating algorithm from point 4.
9. Use the already trained network on-line and generate the desired process variable predictions.

5 Description of the Case Studies

The learning, recall, and generalization characteristics of the neural networks for state estimation in dynamic bioprocesses were investigated using two case studies. Both are based on the ethanol production by the anaerobic bacteria *Zymomonas mobilis*. *Z. mobilis* has a number of interesting properties relevant to efficient ethanol production, including tolerance to high substrate and ethanol concentrations and greater productivity than industrially useful yeasts. Given the numerous advantages of *Z. mobilis* for industrial ethanol production, this bioprocess was chosen to illustrate the application of neural network estimators to provide a fast process evaluation. The ability of the neural networks in using historical databases will prove useful in encoding "knowledge" of the process into the neural estimator.

The first case study consists of the batch cultivation of *Zymomonas mobilis*. Previous research in control and optimization of ethanol production by *Z. mobilis* revealed the importance of cultivation temperature for optimum ethanol yields [31]. Several sets of batch cultivation data were obtained at different temperatures, providing a suitable source of information for training the neural network on the bioprocess behavior at various environmental conditions. This case study illustrates the use of an experimental historical database for training the neural network in the bioprocess behavior.

The second case study involves the simulation of the *Z. mobilis* cultivation in a cell-recycle system using an unstructured model. This cell-recycle operation was evaluated with a view to develop a process with high productivity and high ethanol yields. High ethanol productivity can be achieved in this system because a chemostat with partial cell-recycle can be operated at dilution rates not limited by the "wash-out" condition. Simulation runs were performed at different time-varying conditions to provide enough training data for the neural networks. Future work will involve the experimental validation of the results obtained in these simulation runs.

The case studies under consideration involve a greater degree of complexity than those presented by McAvoy et al. [23], Hoskins and Himmelblau [24], Venkatasubramanian et al. [3] and Bhat and MacAvoy [5]. The case studies presented by these researchers involved chemical or biochemical processes operating at steady state subjected to noise and disturbances or involved fault-diagnosis problems. These studies represent static "pattern" recognition problems where the neural network can be applied as a classifier. In our study, the neural networks are tested using highly nonlinear and continuously time-varying systems.

6 Case Study I. Experimental Batch Cultivation

6.1 Materials and Methods

The *Zymomonas mobilis* bacterial cultivation for ethanol production proceeds anaerobically, where glucose or fructose are metabolized to ethanol, carbon dioxide and lactic acid, and can be run in batch or continuous mode. For our initial experimental evaluation of the neural network estimator, batch runs at different environmental conditions were used.

6.1.1 Organism

Zymomonas mobilis, strain ATCC 10988, was used in this study. It was obtained as lyophilized pellets from the American Type Culture Collection (Rockville, MD).

6.1.2 Medium

Cultures were maintained by weekly transfer in minimal growth media containing 1.0 g l^{-1} yeast extract, 0.5 g l^{-1} Bacto-peptone, and 2.0 weight percent glucose [31]. The liquid cultures were incubated anaerobically under an 85% nitrogen, 5% carbon dioxide, and 10% hydrogen atmosphere at 30 °C. Seed cultures were scaled-up with 5 weight percent glucose, 1.0 g l^{-1} $(NH_4)_2SO_4$, 1.0 g l^{-1} KH_2PO_4, 0.5 g l^{-1} $MgSO_4 \cdot 7 H_2O$, and 3.0 g l^{-1} yeast extract. The glucose and mineral salts were heat sterilized separately for 20 min at 121 °C, and mixed aseptically.

6.1.3 Batch Culture System

Cultivations were conducted in a 7-liter Chemap bioreactor with a working volume of 3 liters, including piping. Prior to inoculation, the entire assembly was heat sterilized at 121 °C for 60 min. Cultures were seeded with 10% volume inoculum, and made anaerobic by continuously sparging N_2 gas. Stirrer speed was set at 150 rpm to provide gentle mixing.

The bioreactor has a built-in pH and temperature controller, and is equipped with sterilizable Ingold pH and redox potential electrodes. The temperature control system consists of a contact J-type thermocouple, a variable set point, a 240 volt/1000 watt heater and a heat exchanger. The pH controller receives the measurement from the pH electrode and operates on a Watson-Marlow peristaltic pump to add 1 N NaOH. The exhaust biogas is passed through the condenser and through a silica gel filter to remove water vapor. The CO_2 gas concentration is measured by means of an Infrared CO_2 Analyzer, and the total gas flow rate is measured by an Omega Mass Flowmeter. All the measurement signals are received by an HP3497A Data Acquisition and Control Unit and processed by an IBM-PC compatible microcomputer for data logging and control.

6.1.4 Analytical Methods

Off-line samples of 30 ml were collected and analyzed for biomass, glucose and ethanol concentrations. Optical density measurements of the culture were determined by measuring the absorption at 575 nm of a 5 ml sample in a Bausch and

Lomb Spectronic 21 spectrophotometer. The remaining 25 ml sample was immediately centrifuged at 18000 rpm for 12 min at 5 °C. The biomass dry weight was
determined from the pellet, and the supernatant was further analyzed for glucose
and ethanol. The biomass pellet was washed three times with distilled water and
dried at 100 °C until constant weight is achieved. The supernatant was filtered
and injected into a Waters HPLC unit with a Bio-Rad Organic Acid analysis
column to quantify the ethanol and glucose concentrations.

6.2 Analysis of the Neural Network Estimator

6.2.1 Training Data and Training Procedure

Five experimental data sets were obtained at temperatures 30 °C, 33 °C, 35 °C,
37 °C and 39.5 °C to provide information of the process behavior at different
environmental conditions. Data was recorded from the available on-line sensors
and from off-line sample analysis performed every 15 minutes. The estimator was
required to predict current biomass, glucose and ethanol concentrations every
15 minutes, using on-line measurements of temperature, redox potential, % CO_2
in exhaust bioprocess gas, and optical density (or an equivalent measurement
thereof, such as turbidity). These four bioprocess variables are assumed to provide
enough information to the neural network about the current state of the process.
It is important to note, though, that the initial species concentrations and initial
cell viability were assumed the same for all data sets. Drastic differences in any
of this factors may cause a deterioration in the neural network estimator performance.

It is well known that the temperature influences the onset and duration of cell
growth and production of metabolites. However, the interpretation of the redox
potential obtained from a bacterial culture is more difficult, because its value is
dependent on more than one factor. It is known that the redox potential value
depends on the temperature, pH and oxygen content of the culture, but more
important, by measuring the reduction-oxidation potential of the medium, a
relationship with the cell viability can be established. The % CO_2 in the biogas

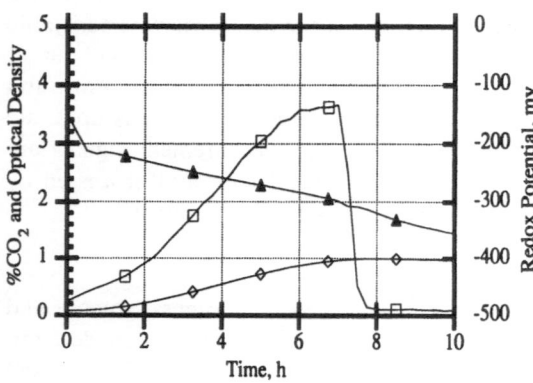

Fig. 5. Inputs for neural network training. Shown are the
on-line measurements for the
T = 37 °C batch data set:
——◇—— Optical density (OD),
——☐—— % CO_2 in exhaust
bioprocess gas, and ——▲——
Redox potential (mv)

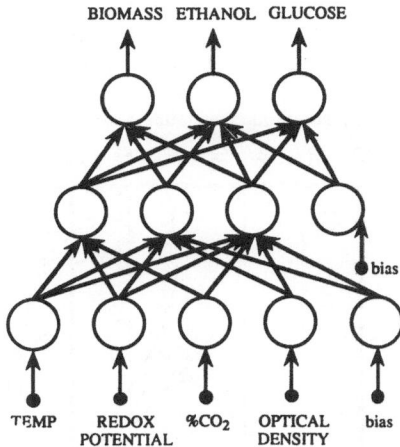

Fig. 6. Proposed neural network configuration for estimation of state variables in ethanol batch production

is an indication of the Carbon Dioxide Evolution Rate (CER), which is directly related to the amount of ethanol produced. Swings and De Ley [32] reported a molar theoretical yield of 1.05 mole CO_2 per mole of ethanol produced. The measurement of the turbidity or optical density is provided as an indirect measure of the cell growth. Figure 5 shows an example of the measurements used as inputs to the net, for the data set at T = 37 °C. The measurements are shown unnormalized. The measurement patterns used to train the network were normalized between 0.1 and 0.9, instead of the usual limits of 0 and 1, because these values were found to improve convergence speed during training.

The neural network employed in this case study has four input nodes corresponding to the four on-line measurements of the system, and three output nodes corresponding to the three state variables to be estimated. A bias node was included in the input layer to provide a threshold for activating the neurons. A prototype network configuration is shown in Fig. 6. The final network architecture will be established in the paragraphs that follow.

One of the tasks in system identification using neural networks is the *training* process. That is, given a training data set, the "best" methodology for presenting the data to the net has to be found, in such a way that the neural network can *learn* the behavior of the system accurately. In this case, it is not possible to join all the data sets to produce one large set of training patterns, due to the time dependency inherent to the bioprocess measurements. Therefore, the training procedure followed was to present the data sets corresponding to different temperatures to the net in a sequential manner. In other words, a training set from one temperature condition was used to train the net until convergence. Then, using those converged weights, the next training set for the next temperature was presented, an so forth. Figure 7 shows a comparison of the experimental values and the neural net estimations for the 5 data sets, using a network configuration of 4 inputs nodes, 4 hidden nodes and 3 output nodes. The order of presentation of the training data to the neural net was found to be of great influence in the

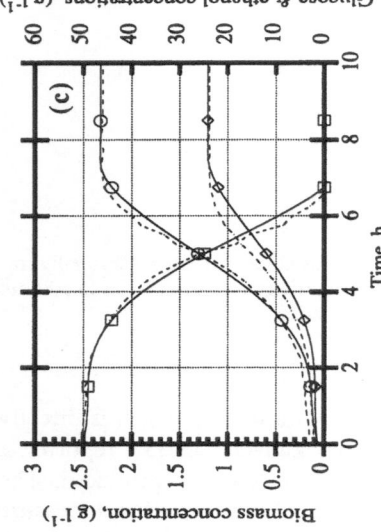

Fig. 7c

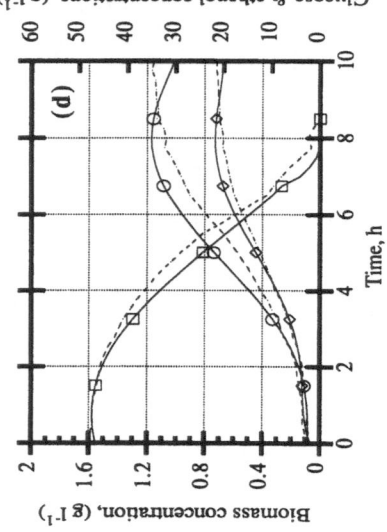

Fig. 7d

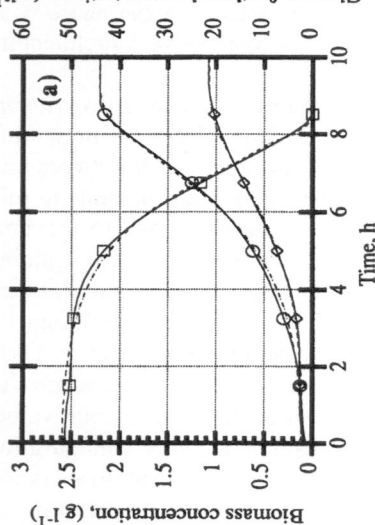

Fig. 7a

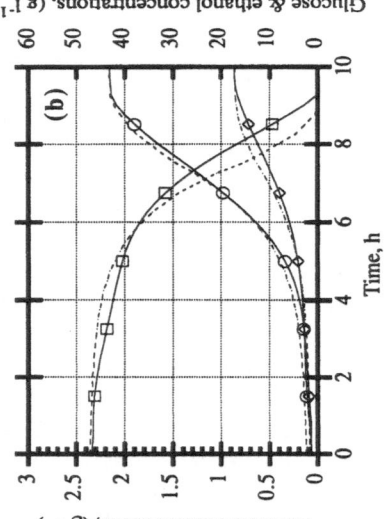

Fig. 7b

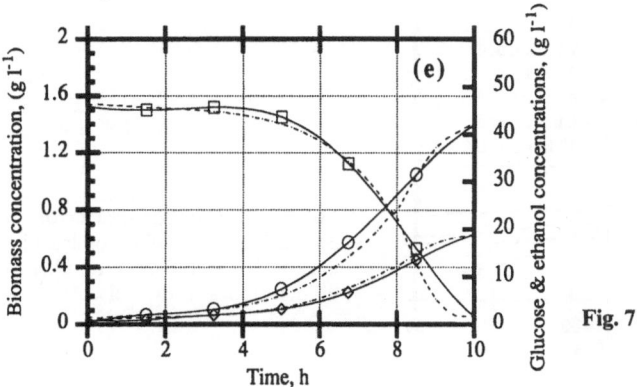

Fig. 7e

Fig. 7. Neural network test data and estimations for all batch data sets using converged network by sequential training. (a) T = 30 °C, SSE = 0.007 (b) T = 33 °C, SSE = 0.12 (c) T = 35 °C, SSE = 0.1 (d) T = 37 °C, SSE = 0.18 (e) T = 39.5 °C, SSE = 0.06. ———— Experimental data, ----------- Estimation. SSE is sum of squared error. —○— Biomass —□— Glucose —◇— Ethanol

estimation accuracy. One can observe that the last pattern presented during training is the one that the neural net recalls better. In Fig. 7 and all the following figures, the training data was presented to the net in a sequential order from T = 39.5 °C to T = 30 °C.

The ability of the neural network to recall patterns it has been trained for and to recognize patterns *it was not trained for* depends in a number of factors, i.e. network architecture, nature and number of data sets, and training procedure. The definition of the network topology is largely based upon a heuristic approach, with some rough guidelines being given by previous researchers. Concerning the network architecture, a sufficient number of layers and neurons per layer has to be defined, in order to approximate the nonlinear process within a given accuracy. The number of neuron units in the input and output layer is basically defined by the nature of the estimation required. In general, the number of hidden layers and hidden neurons is not known. Cybenko [33] has proved that *at most* two hidden layers are required for any given system, with arbitrary accuracy being obtainable given enough units per layer. Moreover, it has been proved that only *one* hidden layer is enough to approximate any *continuous* nonlinear function.

The influence of varying hidden nodes in the network topology was investigated by analyzing the estimation error. Figure 8 shows how the value of the error function (in this case, the sum of squared errors in the estimation, SSE) changes as training proceeds. Plots are shown for 2, 3, 4, 6 and 10 hidden nodes. The error function trajectory is typical of any neural network training process, where there is a rapid decline in error during the first few iterations as the weights are improved from random values to "best-fit" values through the optimization procedure. As more iterations are performed, the error function reaches a *plateau*, where there is little improvement in the optimization. From Fig. 8 it can be seen that a network with three hidden nodes provides the lowest estimation error, and therefore, three

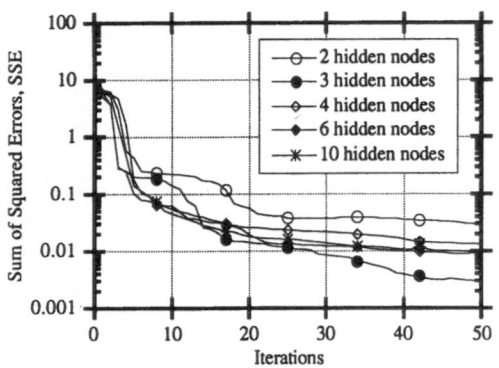

Fig. 8. Sum of squared errors in estimation vs number of learning iterations for different number of hidden nodes in the network

hidden nodes were used in this study. It is to be noted that the computation time per iteration increases exponentially as the number of hidden nodes increases, and therefore, from a computational cost point of view, it is also recommended to use the lowest number of hidden nodes possible. Figure 9 shows the average time in seconds required to perform 50 iterations for different number of hidden nodes.

The calculations were carried out in a Digital Systems DS3100 workstation using Le's Conjugate Gradient Optimization algorithm [30] for training. This conjugate gradient optimization algorithm reduces the number of iterations required for convergence, compared with the ones required using normal steepest descent. The size of the random weights used to initialize the network are found to have a major influence on the convergence ability of the algorithm. If the weights are too large the sigmoid function will saturate from the beginning, and the system will become stuck in a local minimum near the starting point [34]. For this study, the weights w_{ij} were taken to be of the order of $1/\sqrt{k_i}$, where k_i is the number of neurons j which feedforward to neuron i. In this way, the magnitude of a typical net input to unit i is less than, but not too much less than unity. Using this approach, out of 100 training runs, 54 converged to a global minimum and 8 converged to a local minimum. Moreover, it was observed that there is no *unique*

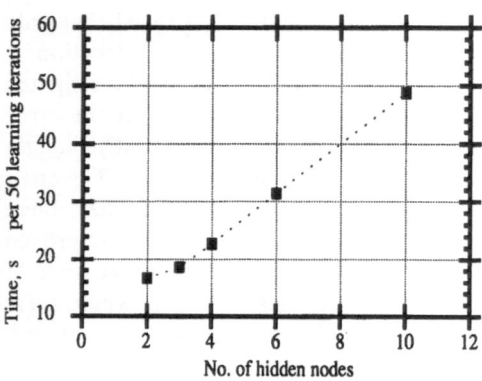

Fig. 9. Average time (s) required for 50 iterations using different number of hidden nodes. The network has 4 input nodes and 3 output nodes. Calculations were performed in a Digital Systems 3100 workstation using Le's Conjugate Gradient optimization algorithm

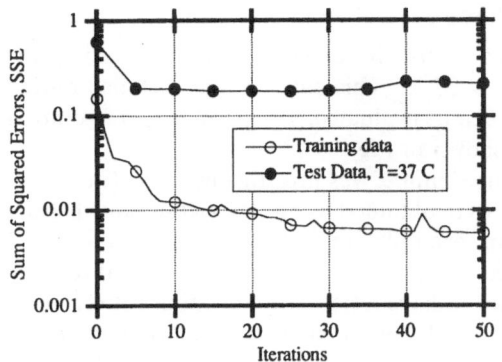

Fig. 10. Sum of squared errors in estimation vs number of iterations for the training data set and the test data set (T = 37 °C)

global minimum, since comparable SSE values were obtained after 50 iterations, but the network weights are not the same, not even in the sign. This suggests a highly convoluted error surface with several minima.

The influence of varying the number of learning iterations was also investigated. Figure 10 shows the error function SSE for the training data and the test data set T = 37 °C. Starting from initially random weights, the training was performed for 50 iterations. At every 5th iteration, the set of weights up to that point was used to evaluate the estimation error with the test data. As the number of learning iterations increases, the ability of the neural network to properly recall the test data decreases. There exists an optimum value of learning iterations, which provides a compromise between the *learning* and *reall* characteristics of the network. This minimum value of SSE is obtained at 25 iterations. The same conclusion can be obtained from a different exercise. Figure 11 shows the SSE during sequential training using 25 learning iterations and 50 iterations for each of the five data sets. It can be seen that 25 learning iterations per training data set provides the least SSE. In other words, a longer training does not necessarily lead to better performance in pattern recall.

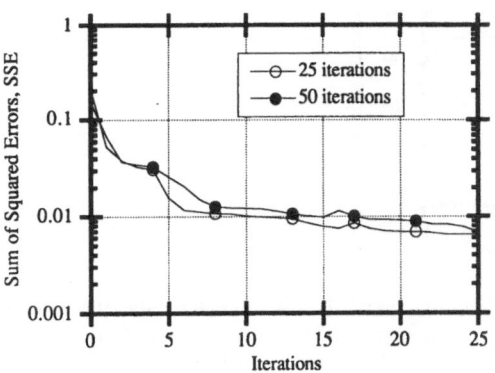

Fig. 11. Sum of squared errors in estimation vs number of iterations during sequential training using 25 iterations and 50 iterations per data set

6.2.2 Test Data and Testing Procedure

The neural network estimator was tested in different situations using the experimental data set obtained at T = 37 °C. This data set was included during training, so the recall characteristics can be evaluated. The network configuration used was defined previously and is shown in Fig. 5.

The influence of noise in the measured inputs was investigated next. The neural network estimations shown in Fig. 7 were obtained using raw measurements from the biosensors, including noise. For example, Fig. 7d shows the data set at T = 37 °C and the neural network estimations. The SSE value is 0.18. The temperature measurements are known to have a standard deviation between 0.01 and 0.03 °C, for the different experimental conditions. If a prefiltering algorithm is included to reduce the noise in the measurements used as inputs to the net, a better estimation is obtained. Figure 12 shows the neural network estimations for the data set at T = 37 °C, assuming there is no or little noise in the temperature measurements. A SSE of 0.08 was obtained in this case. Comparing these error function values, it can be seen that the error is more than twice as much in the case where noise is present. In our case, the error is not large enough to justify a prefiltering of the input data. In other situations, it might be necessary to prefilter the input data, since the neural network estimation may deteriorate in the presence of large amounts of noise.

Other researchers [35] have found that for fed-batch and continuous cultivations 3 network inputs are pertinent to provide enough information for training, these being time, CO_2 evolution rate, and substrate feed. The influence of time as an input to the net has not been reported for batch cultivations. Figure 13 shows estimations for the batch data set with T = 37 °C when the time is included as an input to the net. The error function value SSE for this case is 0.088, whereas the SSE obtained without the time input is 0.18. This suggests that the time dependency is important, especially in batch systems. Therefore, the temporal

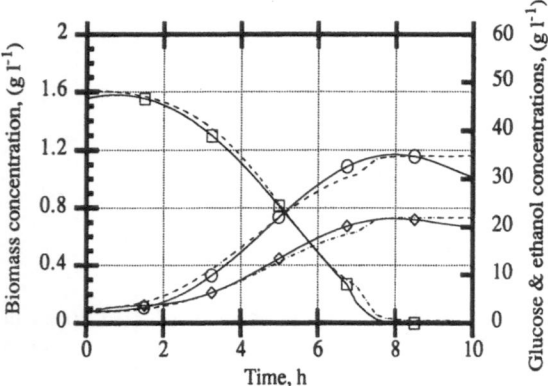

Fig. 12. Neural network test data and estimations for the T = 37 °C batch data set assuming no noise is present in the temperature measurements. SSE = 0.08. ——— Experimental data, ------------ Estimation. —⊖— Biomass —⊟— Glucose —◇— Ethanol

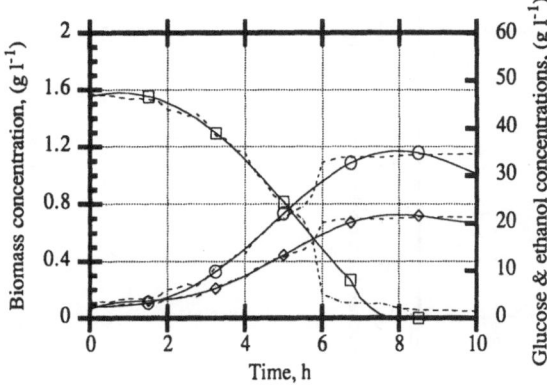

Fig. 14. Neural network test data and estimations for the T = 37 °C batch data set including glucose concentrations as an input to the net. SSE = 0.092. ──── Experimental data, ------------ Estimation. ──○── Biomass ──▭── Glucose ──◇── Ethanol

information contained in the sequence cannot be disregarded. Future research should involve the use of time-dependent recurrent networks as estimators for nonlinear dynamic systems.

The influence of including the glucose concentration as an input to the network was next studied. Fig. 14 shows the estimation results for the data set obtained with T = 37 °C. Comparing the SSE obtained here (0.092) with the SSE of Fig. 7d (0.18), we can see that glucose concentration provides additional information on the process, which generates an improvement on the estimation. However, including glucose concentration as an input to the net in this way, is not realistic, since we are assuming that this value is going to be available on-line, and this is not usually the case. If we assume that glucose concentration is only available from an off-line analysis performed every 2 h, and we try to keep the same converged network, the estimation performance deteriorates dramatically. However, improvements in the estimation can be expected if the network is trained from the beginning with infrequent substrate concentrations. In that case, the network is *taught* to expect substrate inputs only after a specified sampling interval, and will *learn* to match this to correct estimated outputs. Figure 15 shows the estimations following

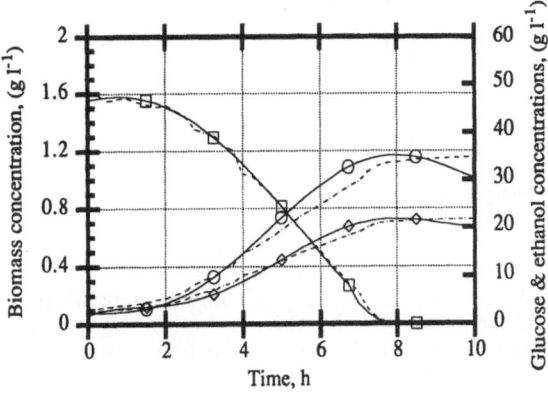

Fig. 13. Neural network test data and estimations for the T = 37 °C batch data set including time as an input to the net. SSE = 0.088. ──── Experimental data, ------------ Estimation. ──○── Biomass ──▭── Glucose ──◇── Ethanol

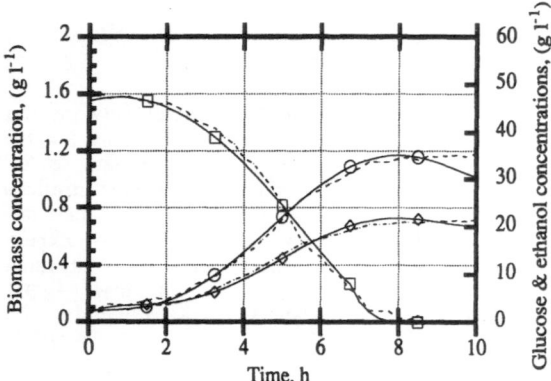

Fig. 15. Neural network test data and estimations for the T = 37 °C batch data set where the training includes infrequent measurements of glucose concentration as an input to the net. Samples are assumed available every 2 h, and values are kept constant in between. SSE = 0.048. ————— Experimental data, ------------ Estimation. —⊖— Biomass —⊟— Glucose —◇— Ethanol

this procedure. An error function value SSE of 0.048 was obtained, almost half as that one obtained previously (SSE = 0.092). The conclusion would be that as much information as is available should be used to train the network. In this way, the neural network is able to provide more accurate estimations.

Neural networks are well-known for their *generalization* characteristics, that is, the ability to predict untrained patterns at conditions other than the ones on which it was trained. The ability of this neural network for generalization was studied using an experimental batch data set obtained with a time-varying temperature profile, which was not used during training. In this batch experiment,

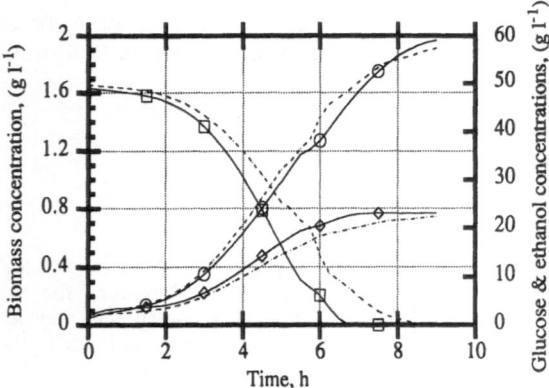

Fig. 16. Neural network test data and estimations for the varying temperature profile batch data set. Temperature profile is 35 °C for the first 5 h, 34 °C for the next hour, and 33 °C for the last 3 h. SSE = 0.23. ————— Experimental data, ------------ Estimation. —⊖— Biomass —⊟— Glucose —◇—Ethanol

the temperature is assumed to vary in a step fashion with T = 35 °C for the first
5 h, then T = 34 °C for the next hour and finally T = 33 °C for the last 3 h. The
converged network of Fig. 7 was used to generate the estimations. The data set
and results are shown in Fig. 16. The estimations for the biomass and ethanol
concentrations are good, but the estimation for glucose concentrations could be
improved. Better estimations can be obtained by incorporating some of the
additional information discussed in previous paragraphs. In general, this study
demonstrates the ability of the network in predicting patterns not presented during
training.

7 Case Study II. Cell-Recycle System Simulation

7.1 Model for Ethanol Production: Cell-Recycle System

The model of growth and product formation of *Z. mobilis* in a recycle reactor is
based on an unstructured model derived by Lee et al. [36]. Though models with
a higher degree of complexity are available [37, 38], an unstructured model was
found to be appropriate for our intended application, that is, process development
and control. This nonlinear process model includes some modifications to make
it consistent with the structured kinetic model developed by Veeramallu and
Agrawal [38].

Kinetic Model Assumptions:

1. A Monod-type kinetic model is assumed for the effect of substrate limitation
   on the specific growth rate and specific ethanol productivity.
2. Nonlinear terms account for biomass, substrate and product inhibition on
   specific growth rate and specific ethanol productivity.
3. No CO_2 inhibition.

   According to these assumptions, the definitions of the specific growth rate μ
and specific ethanol production rate q_p are as follows:

$$\mu = \mu_{mo} \frac{s}{K_s + s} \left[1 - \left(\frac{p}{P_m} \right)^a \right] \left[1 - \frac{x}{x_{\max}} \right]^2 \frac{(K_i - S_i)}{(K_i - S_i) + (s - S_i)} \quad (20)$$

$$q_p = q_{pm} \frac{s}{K'_s + s} \left[1 - \left(\frac{p - P'_i}{P'_m - P'_i} \right)^b \right] \left[\frac{K'_i}{K'_i + s} \right] \quad (21)$$

where $(s - S_i) = 0$ for $s \le S_i$ and $(p - P'_i) = 0$ for $p \le P'_i$. Definitions of the
biological constants and their approximate value are listed in Table 1. These values
have been obtained from the literature for a constant temperature of 30 °C and
pH = 5. Any change in these environmental conditions will affect the value of
the kinetic parameters. The pH value is usually kept constant during the course
of the cultivation, but temperature may vary. If temperature variations need to
be considered, the maximum specific growth rate can be expressed as an Arrhenius
function of temperature. Two different Arrhenius functions can be specified: one for

Table 1. Biological constants for *Zymomonas mobilis* kinetic model

Parameters	Descriptions	Value
μ_{mo}	Max specific growth rate at zero ethanol concentration	$0.4\,\mathrm{h}^{-1}$
K_s	Monod kinetic constant	$0.5\,\mathrm{g\,l}^{-1}$
P_m	Maximum ethanol concentration for cell growth	$217\,\mathrm{g\,l}^{-1}$
a	Power of the ethanol inhibition term	0.64
x_{max}	Maximum cell concentration at dense packing	$80\,\mathrm{g\,l}^{-1}$
K_i	Substrate inhibition constant for growth	$200\,\mathrm{g\,l}^{-1}$
S_i	Threshold substrate concentration for cell growth	$80\,\mathrm{g\,l}^{-1}$
q_{pm}	Maximum specific ethanol production rate	$4.26\,\mathrm{g\,g}^{-1}\,\mathrm{h}^{-1}$
K'_s	Saturation constant for q_p	$0.5\,\mathrm{h}^{-1}$
P'_m	Max ethanol concentration for ethanol production	$110\,\mathrm{g\,l}^{-1}$
P'_i	Ethanol threshold concentration for ethanol prod'n.	$50\,\mathrm{g\,l}$
b	Power of the ethanol inhibition term in q_p	0.47
K'_i	Substrate inhibition constant for ethanol production	$500\,\mathrm{g\,l}^{-1}$
$Y_{p/s}$	Ethanol yield	$0.47\,\mathrm{g\,g}^{-1}$

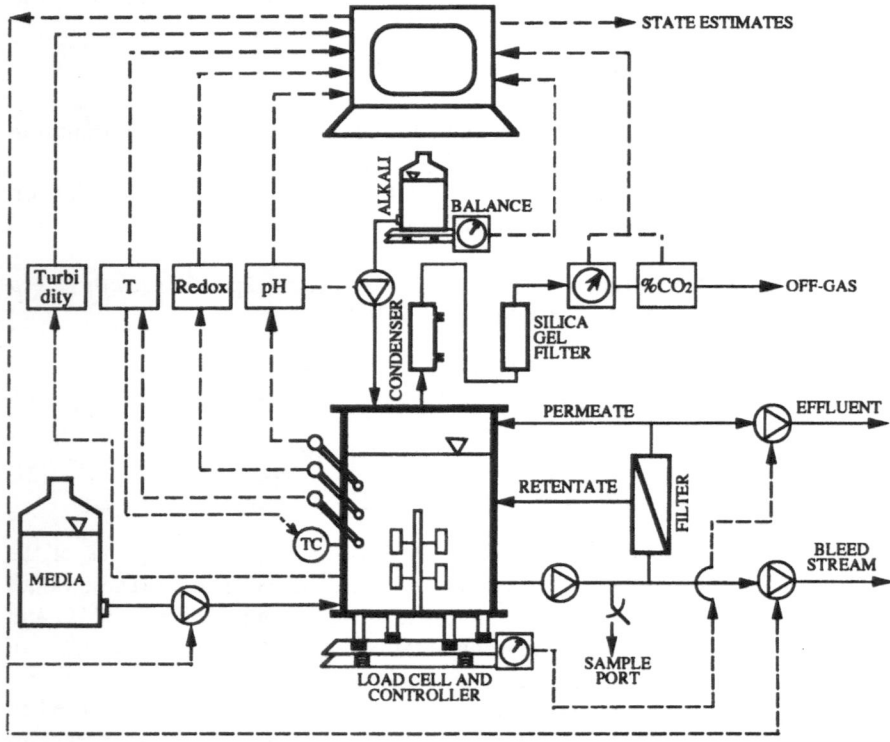

Fig. 17. Experimental set-up for *Zymomonas mobilis* cell-recycle operation

the activation phase and another for the deactivation phase. In this way, the equation will be valid for the temperature range considered.

A schematic of the chemostat with cell-recycle is shown in Fig. 17. Assuming sterile feed with no ethanol content, a constant yield coefficient, and a completely mixed, constant volume bioreactor with continuous loading and cell recycling, the unsteady state mass balances yield:

$$\frac{dx}{dt} = \mu x - D(1 - R) x \tag{22}$$

$$\frac{dp}{dt} = q_p x - DRp \tag{23}$$

$$\frac{ds}{dt} = -\frac{1}{Y_{p/s}} q_p x + D(s_0 - Rs) \tag{24}$$

where x, p and s are the biomass, ethanol and glucose concentrations in the bioreactor $(g\,l^{-1})$, respectively. Also, D is the dilution rate (h^{-1}), R is the recycle ratio (F/F_0), F is the permeate flow $(l\,h^{-1})$, F_0 is the total feed flow rate to the system $(l\,h^{-1})$ and s_0 is the glucose concentration in the feed flow $(g\,l^{-1})$.

7.2 Simulation Studies and Results

The purpose of the simulation studies is to gain experience and evaluate the performance of the neural network estimator in continuous systems. Using the process model developed in the previous section, a set of data was generated to be used as a training set or *teacher signal* for the neural network estimator. The case study analyzed, assumes that biomass concentration (or an equivalent on-line measurement) and dilution rate are available as inputs to the net, and a prediction of the glucose and ethanol concentration is required. This would correspond to a two-input-two-output case in identification. In an experimental set-up, a variable which correlates well with biomass concentration, such as optical density, turbidity, etc., can be used instead of the assumed available biomass.

The values used in the simulations are as follows: recycle ratio $R = 0.99$, mean dilution rate $D = 10\,h^{-1}$ and mean feed substrate concentration $s_0 = 100\,g\,l^{-1}$. Corresponding to these values, the steady-state values of biomass, glucose and ethanol concentrations are $39.76\,g\,l^{-1}$, $69.27\,g\,l^{-1}$, and $14.91\,g\,l^{-1}$, respectively.

Figure 18 shows the training data used for two inputs to the net, these being biomass concentration and dilution rate. The dilution rate varied in square wave fashion with random amplitude value, the minimum and the maximum value being $10\,h^{-1}$ and $14\,h^{-1}$, respectively. Further, a 5% amplitude random noise signal was superimposed to observe the effect of possible measurement noise in the estimation. The biomass measurement was not altered in any way, and the fluctuations it shows, are due to the coupled effect with dilution rate variations. The simulation was run for 50 h generating 250 sampled values throughout. These values were used as training data.

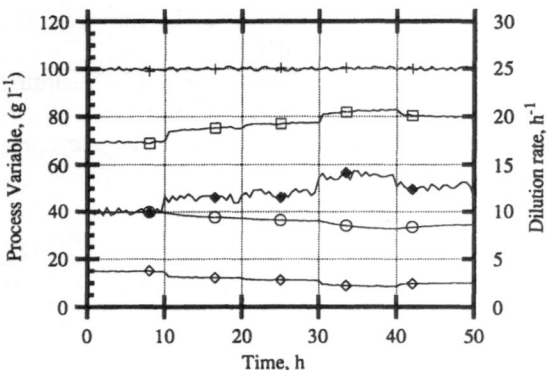

Fig. 18. Training data for the 2-input neural network. Inputs are —⊖— biomass concentration (g l^{-1}) and —◆— dilution rate (h^{-1}). Outputs are —▢— glucose and —◇— ethanol concentrations (g l^{-1}). Feed substrate —+— So (g l^{-1}) is provided as a disturbance in the process

In order to test the neural network estimator, the dilution rate was varied following a triangular wave pattern to investigate the ability of the neural net to predict patterns it was not trained for. The test data used is shown in Fig. 19. Results for different network topologies can be observed in Fig. 20. From this figure it can be seen that as the number of hidden nodes increases, the estimation accuracy deteriorates. This is because higher order neural nets are able to map higher order nonlinearities, and in fact, may act like noise amplifiers, as was discussed in Sect. 6.2.2. In some cases, prefiltering would be necessary to reduce the amplitude of the existent noise in the measurements. The network configuration used in subsequent studies contains 2 inputs nodes, 2 hidden nodes and 2 output nodes.

Finally, the neural network estimator was tested with a data set where the maximum specific growth rate changes. This was done to simulate the ability of

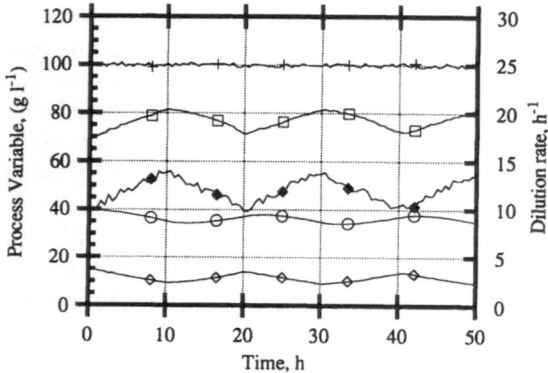

Fig. 19. Test data for the neural network estimator. See Fig. 18 for definition of variables

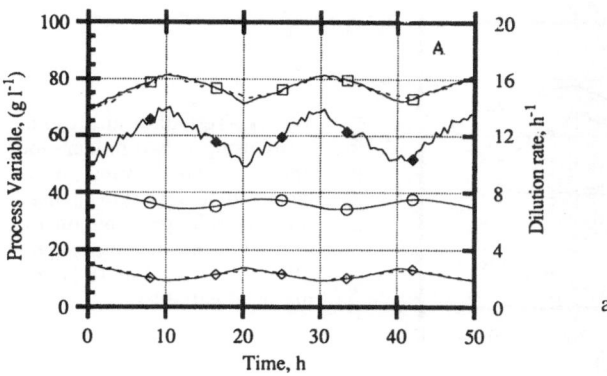

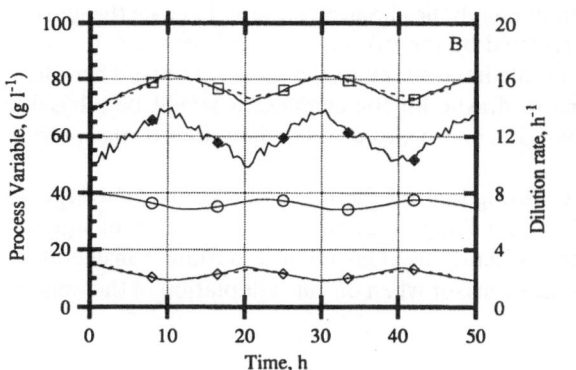

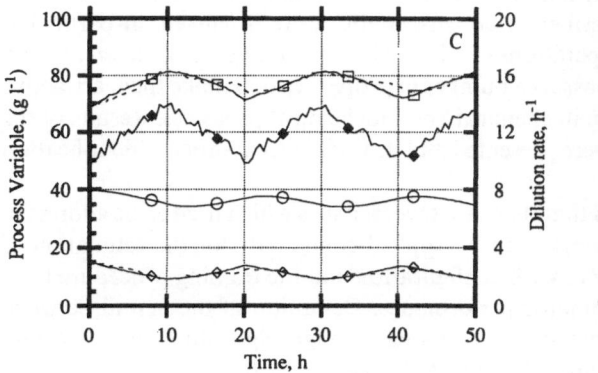

Fig. 20a–c. Comparison of estimator performance using different neural network topologies. (a) 2 hidden nodes, SSE = 0.376; (b) 3 hidden nodes, SSE = 0.393; (c) 4 hidden nodes. SSE = 1.024. See Fig. 18 for definition of variables

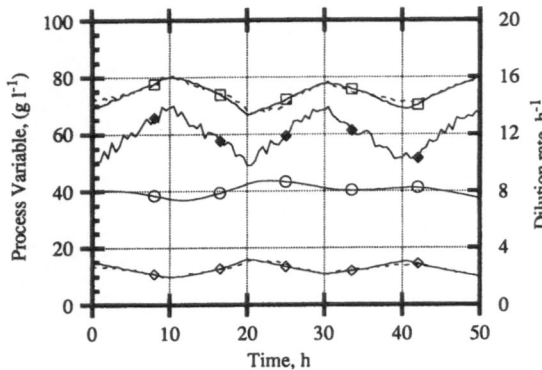

Fig. 21. Test data and estimations using a two-hidden-node neural network estimator. Includes 20% variation in maximum specific growth rate μ_{max} around the mean value 0.45 h^{-1} SSE = 0.47. See Fig. 18 for definition of variables

the estimator to handle process kinetics variations, which frequently happen in biological systems due to different metabolic changes. Figure 21 shows the glucose and ethanol concentrations predicted by the two-input neural network. It can be seen that the results are comparable in accuracy to the ones obtained when there are no kinetic changes. For more drastic kinetic changes, it would be advisable to adjust the neural network weights on-line, to enable adaptation to the varying process conditions.

It is worth mentioning that, throughout all the training sessions, convergence speed was found to be very high, reaching an almost constant value of squared error (SSE) in between 10 and 20 iterations. This is an important consideration when working with on-line applications, or when on-line adaptation of the weights has to be performed.

8 Conclusions

The application of artificial neural networks for estimation of state variables in bioprocesses has been described. Two case studies were discussed, in particular, the batch and cell-recycle operations of ethanol production by *Zymomonas mobilis*. Artificial neural networks possess a number of properties that make them attractive for state variable estimation in chemical and biochemical processes. Results of the neural network estimator were presented and its usefulness in process identification was demonstrated.

However, it is to be noted that there are several issues which need to be addressed before artificial neural networks can be applied confidently for the estimation of state variables. The neural network configuration and the training procedure have major influences in the estimator performance. Definition of the neural network configuration still relies heavily on heuristics, though algorithms have recently been developed which dynamically adjust themselves to the requirements of the training process [39]. The network topology, meaning number of inputs nodes, number of hidden nodes and number of output nodes, is of great importance, since this will define the network's abilities for learning and generalization.

The training procedure involves not only the methodology for presenting the inputs to the net, but also how to form an adequate training signal. A training data set should be large enough to provide enough information on the process, but not so large that it is computationally time consuming to implement in real-time. Furthermore, the training signal should provide enough information about the process in order to span the state-space area of interest. Data conditioning should be considered in the cases when there is a considerable amount of noise in the measurements, otherwise deterioration in the prediction may occur. Several training procedures have been investigated, and one cannot define a single training method that will apply for all the cases. It is highly dependent on the nature of the process to be identified.

Recently, there has been a lot of research in developing methods for faster learning. In the cases where adaptation has to be performed on-line, a fast optimization algorithm for converging the network weights is required. The algorithm presented here converged fast in all the cases that were analyzed. As it is well-recognized, it is more difficult to apply artificial neural networks to time-varying problems than to static pattern recognition, such as fault diagnosis. Thus, if fast learning algorithms are available, on-line adaptation can be carried out to improve performance.

In the cases discussed in this paper, it was assumed that enough information about the process was provided in the teacher signal. Also, the bioprocess initial conditions were assumed to be the same for all the data sets processed, including training and testing sets. The neural network estimator can also be used as a diagnostic tool. If the bioprocess is assumed to be under normal operation, the estimated values would be following the usual trend. If the bioprocess is under faulty conditions, such as contamination, the neural network estimates will deviate noticeably from the normal conditions and one could conclude that there is a problem with the cultivation, and subsequently take corrective actions.

In summary, an artificial neural network trained by backpropagation was found to be useful for bioprocess identification and diagnosis. The estimator techniques described in this paper have been shown to provide acceptable on-line process estimates. The incorporation of environmental conditions as inputs to the net was shown to provide additional useful information to aid the estimation.

Acknowledgements: This work was partly funded by the Colorado Institute for Research in Biotechnology (CIRB) and the Colorado State University Experimental Station.

9 References

1. Ungar LH (1990) A bioreactor benchmark for adaptive network-based process control. In: Sutton Miller RS, Werbos WT, Werbos PJ (eds) Neural Networks for Control. MIT Press, 387
2. Ungar LH, Powell BA, Kamens SN (1990) Computers in Chemical Engineering 14: 561

3. Venkatasubramanian V, Vaidyanathan R, Yamamoto Y (1990) Computers in Chemical Engineering 14: 699
4. Vaidyanathan R, Venkatasubramanian V (1990) Representing and diagnosing process trends using neural networks. In AIChE Annual Meeting, Chicago, Il, November 1990
5. Bhat N, McAvoy TJ (1990) Computers in Chemical Engineering, 14: 573
6. Rumelhart DE, Hinton GE, Williams RJ (1986) A general framework for parallel distributed processing. In: McClelland JL, Rumelhart DE, the PDP Research Group (eds) Parallel Distributed Processing: Explorations in the Microstructure of Cognition. MIT Press p 45
7. Rogers PL, Lee KJ, Skotnicki ML, Tribe DE (1982) Advances in Biochemical Engineering 23: 37
8. Carleysmith SW, Fox RI (1984) Advances in Biotechnological Processes 3: 1
9. Tenorio MF (1990) IEEE Transactions on Neural Networks 1: 100
10. Wang HY., Cooney CL, Wang DIC (1979) Biotechnology and Bioengineering 21: 975
11. Zabriskie DW, Humphrey AE (1978) AIChE Journal 24: 138
12. Wang NS, Stephanopoulos GN (1984) CRC Critical Reviews in Biotechnology 2: 1
13. Rolf MJ, Lim HC (1985) Biotechnology and Bioengineering 27: 1236
14. Andersen MY, Jorgensen SB (1989) Identification of a simulated continuous yeast fermentation. In: Fox RI, Fish NM, Thornhill NF (eds) Computer Applications in Fermentation Technology: Modelling and Control of Biotechnological Processes. Elsevier, Amsterdam 205
15. Petersen JN, Whyatt GA (1990) Biotechnology and Bioengineering 35: 712
16. Stephanopoulos G, San K-Y (1984) Biotechnology and Bioengineering 26: 1176
17. Montgomery PA, Williams D (1989) On-line estimation of cell mass using an Extended Kalman Filter. In: Fox RI, Fish NM, Thornhill NF (eds) Computer Applications in Fermentation Technology: Modelling and Control of Biotechnological Processes. Elsevier, Amsterdam, 221
18. Shimizu H, Takamatsu T, Shioya S, Suga K-I (1989) Biotechnology and Bioengineering 33: 354
19. Ryoo D, Murphy VG, Karim MN, Tengerdy RP (1991) Biotechnology and Bioengineering. In press.
20. Bastin G, Dochain D (1986) Automatica 22: 705
21. Pomerleau Y, Perrier M (1989) Nonlinear estimation and adaptive control of a fed-batch fermentor. In: Fox RI, Fish NM, Thornhill NF (eds) Computer Applications in Fermentation Technology: Modelling and Control of Biotechnological Processes. Elsevier, Amsterdam, 361
22. Qian N, Sejnowski TJ (1988) Journal of Molecular Biology 202: 865
23. McAvoy TJ, Wang NS, Naidu S, Bhat N, Gunter J, Simmons M (1989) Interpreting biosensor data via backpropagation. In: IJCNN International Joint Conference on Neural Networks, San Diego, CA, vol 1 p 227
24. Hoskins JC, Himmelblau DM (1988) Computers in Chemical Engineering 12: 881
25. Haesloop D (1990) A neural network structure for system identification. In: Proceedings of the 1990 American Control Conference, May 23–25 p 2460
26. Breusegem VV, Thibault J, Cheruy A (1991) Canadian Journal of Chemical Engineering 69: 481
27. Thibault J, Breusegem VV, Cheruy A (1990) Biotechnology and Bioengineering 36: 1041
28. MacGregor JF, Marlin TE, Kresta JV (1991) Some comments of neural networks and other empirical modelling methods. In: Arkun Y, Ray WH (eds) Chemical Process Control CPC-IV, CACHE and AIChE, New York, p 665
29. Leonard J, Kramer MA (1990) Computers in Chemical Engineering 14: 337
30. Le D (1985) Mathematical Programming 32: 41
31. Rivera SL, Karim MN (1990) Application of dynamic programming for fermentative ethanol production by Zymomonas mobilis. In: Proceedings of the 1990 American Control Conference, San Diego, CA, May 23–25, p 2144
32. Swings J, De Ley J (1977) Bacteriological Reviews 41: 1
33. Cybenko G (1989) Mathematics of Control, Signal and Systems 2: 303

34. Hertz J, Krogh A, Palmer RG (1991) Introduction to the Theory of Neural Computation, vol. I. Addison-Wesley, Redwood City, CA
35. Morris AJ, Montague GA, Tham MT, Aynsley M, Di Massimo C, Lant P (1991) Towards improved process supervision-algorithms and knowledge based systems. In: Arkun Y, Ray WH (eds) Chemical Process Control CPC-IV, CACHE and AIChE, New York, p 585
36. Lee KJ, Rogers PL (1983) The Chemical Engineering Journal 27: B31
37. Jeong JW, Snay J, Ataai MM (1990) Biotechnology and Bioengineering 35: 160
38. Veeramallu U, Agrawal P (1990) Biotechnology and Bioengineering 36: 694
39. Fahlman SE, Lebiere C (1990) The cascade-correlation learning architecture. In: Touretzky D (ed) Advances in Neural Information Processing Systems 2, Morgan Kauffman, San Mateo, p 524

Use of Regulated Secretion in Protein Production From Animal Cells: An Overview

G. E. Grampp[1], A. Sambanis[2] and G. N. Stephanopoulos[1]

[1] Department of Chemical Engineering and Biotechnology Process Engineering Center, Massachusetts Institute of Technology, Cambridge, MA 02139, USA

[2] School of Chemical Engineering, Georgia Institute of Technology, Atlanta, GA 30332-0100, USA

Traditional industrial cell culture processes require extensive downstream product refining due to low product titer and purity in the spent growth medium. A controlled secretion process incorporating cells derived from endocrine or exocrine organs could potentially alleviate this processing burden by dynamically decoupling product recovery from cell growth and product biosynthesis. In addition, such specialized secretory cells may be uniquely capable of performing desirable post-translational processing of the secretory product. We briefly review the biology of regulated protein secretion as well as the biology and biochemistry of the signal transduction mechanims by which regulated systems respond to environmental stimuli. Drawing on these and other basic principles from cell biology and bioengineering, we describe the important features of a controlled secretion process. Among other issues we discuss the choice of cell lines, expression systems, cell culture methods, and bioreactor configurations. We extensively analyze the kinetics of regulated secretion in the context of a controlled secretion process. This discussion is illustrated with experimental results from two model cell lines, recombinant AtT-20 and βTC3, expressing recombinant human endocrine hormones or native murine insulin respectively.

Advances in Biochemical Engineering
Biotechnology, Vol. 46
Managing Editor: A. Fiechter
© Springer-Verlag Berlin Heidelberg 1992

List of Abbrevations

cAMP   cyclic adenosine monophosphate
ATP   adenosine triphosphate
Ca^{++} divalent calcium ion
IBMX   isobutylmethylxanthine
BrcAMP   8-bromo-cyclic adenosine monophosphate
hGH   human growth hormone
IgG   immunoglobulin G
BiP   immunoglobulin binding protein
ACTH   adrenocorticotropic hormone
DNA   desoxyribonucleic acid
mRNA   messenger ribonucleic acid
DMEM   Dulbecco's modified Eagle's medium
tPA   tissue plaminogen activator

1 Introduction

The structure and function of polypeptide products derived from mammalian sources are often dependent upon a number of post-translational modifications inlcuding enzyme-catalyzed folding, glycosylation, proteolytic processing, and several other covalent alterations. Due to their native ability to perform these processes, animal cells are indispensable for the production of many pharmaceutically important proteins and polypeptides. Although the protein of interest may be expressed endogenously in a mammalian cell line, in most (if not all) industrial applications it is expressed in an appropriate host cell by means of a recombinant vector. After the genetic manipulations are completed in the lab, the process is scaled up. Medium formulation is done at an intermediate level, and the final production size is achieved in appropriately designed and controlled bioreactors.

In all industrial processes that we are aware, of, the product protein is secreted constitutively (as it is synthesized) into the cell culture medium. Constitutive secretion imposes a burden upon downstream processing due to the low purity and titer of protein product which result from factors inherent to mammalian cell culture. It may be possible to relieve this burden by using a controlled secretion process with lines of properly differentiated animal cells. With this approach it is feasible to recover the product in medium free of contamination from exogenous proteins and at titers at least 5 to 10 times higher than those achieved using constitutive secretion. As it is also explained later, controlled secretion processes may offer significant advantages with regard to the biological activity of product, especially polypeptide hormones, relative to traditional operations involving constitutive secretion.

The objective of this paper is to provide a comprehensive description of the essential features of a controlled secretion process. This process relies primarily on a biological phenomenon, regulated protein secretion. Therefore, much of the paper is devoted to describing the relevant biological properties of host cells, including cell differentiation, protein expression, post-translational processing, and secretory dynamics. Although specific biological properties and preliminary evaluations of cycling processes have been presented or published, an integrated view that relates cell characteristics, culture conditions, and engineering issues, is essential for assessing the potential of the proposed technology. When possible, ideas relevant to each of these topics are illustrated with reference to experimental results from our laboratory.

1.1 The Biology of Regulated Protein Secretion

Specialized animal cells of endocrine and exocrine glands have the ability to store secretory proteins in intracellular compartments and release them efficiently only when the level of a cytosolic messenger, usually Ca^{++} or cAMP, is altered [1, 2]. This secretory pathway is denoted as regulated (Fig. 1). Proteins secreted in a regulated fashion are sorted at the trans Golgi into secretory vesicles which, after a maturation process, become storage vesicles or granules and accumulate in the post-Golgi region of the cytoplasm [3, 4, 5]. In unstimulated cells, the lifetime of storage granules can be many hours to days. In contrast, constitutive transport vesicles remove proteins from the Golgi and fuse to the plasma membrane after a short (of the order of 10 minutes) residence time in the cytoplasm [1, 2]. A significant advantage of the regulated pathway is that secretion can be decoupled from protein synthesis and controlled for improved product recovery. Furthermore, regulated secretion may be needed to ensure biological activity of the secreted protein. During maturation of regulated secretory vesicles, post-translational processes may occur that include proteolytic cleavage, amidation and acetylation [6]. These modifications can be essential for biological activity of the secreted product, and they do not normally occur in the context of constitutive secretion. Regulated secretory proteins, especially polypeptide hormones, missorted into the constitutive pathway may thus be secreted as biologically inactive molecules. On the other hand, it is possible that constitutive secretory proteins directed into the regulated pathway (by methods described below) may undergo inappropriate post-Golgi modifications. This is an unlikely possibility, however, due to the specificity of the enzymes in regulated secretory vesicles.

The stimulation of the regulated secretion pathway by external signals (secretagogues) is mediated by complex signal transduction mechanisms. Several types of molecules can be involved, including membrane-bound receptors, transduction proteins, effector enzymes, kinases, ion channels, ionophores, secondary and even tertiary messengers. These components may interact in complicated ways to produce synergistic or antagonistic effects [7]. In recent years research on signal transduction has advanced to the point that many of the fundamental mechanisms have been elucidated. While a comprehensive dynamic model of stimulus-response

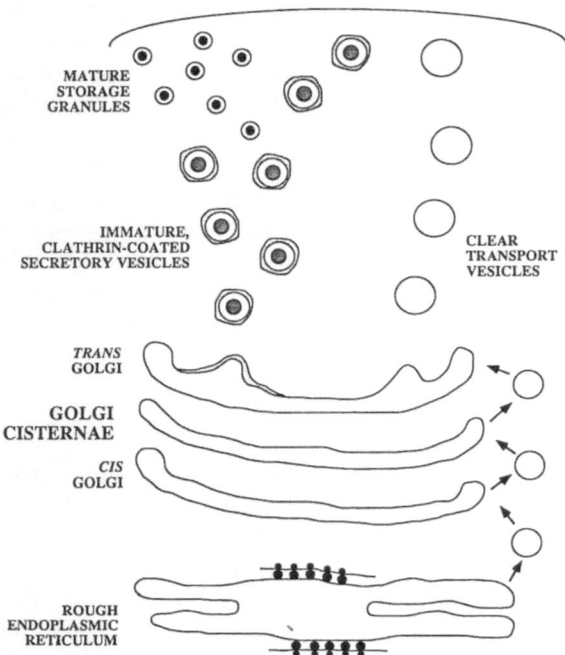

MATURE
STORAGE
GRANULES

IMMATURE,
CLATHRIN-COATED
SECRETORY VESICLES

CLEAR
TRANSPORT
VESICLES

TRANS
GOLGI

GOLGI
CISTERNAE

CIS
GOLGI

ROUGH
ENDOPLASMIC
RETICULUM

Fig. 1. (from 39). Secretory protein trafficking in animal cells. Secretory proteins are synthesized on ribosomes bound on the membrane of the rough endoplasmic reticulum (RER). Proteins are cotranslationally translocated into the lumen of the RER, and they are transported in membrane bound vesicles to the cis Golgi and, through medial Golgi, to the trans Golgi cisternae. These transport steps appear to occur via a bulk flow mechanism. In the RER and Golgi, proteins undergo various posttranslational modifications, such as folding, glycosylation and proteolysis. At the trans Golgi, secretory proteins are sorted into vesicles that take them quickly or after some additional events to the plasma membrane for exocytosis. In most animal cells, secretory proteins enter (apparently by bulk flow) transport vesicles that take them continuously to the plasma membrane (constitutive secretion). In specialized cells of the endocrine and exocrine systems, proteins are actively sorted (by means of "sorting" or "hormonal binding" proteins) into immature, clathrin-coated secretory vesicles that bud off from specialized regions of the trans Golgi. These vesicles undergo a maturation process to become storage vesicles or granules that accumulate in the cytoplasm and fuse with the plasma membrane at a significant rate only when the cells are triggered with an appropriate secretion agonist (regulated secretion). Additional modifications, quite important for biological activity, occur in maturing secretory vesicles

phenomena is not yet feasible, the new developments provide an excellent foundation for detailed studies with specific cell lines.

The basic mechanisms of increase of cytosolic Ca^{++} and cAMP concentrations are illustrated in Fig. 2. Cytosolic calcium can increase through receptor-mediated release of inositol trisphosphate (IP_3), which in turn stimulates release of Ca^{++} from the endoplasmic reticulum storage [8]; through membrane depolarization by electrical or chemical stimulus, followed by influx of extracellular Ca^{++} through

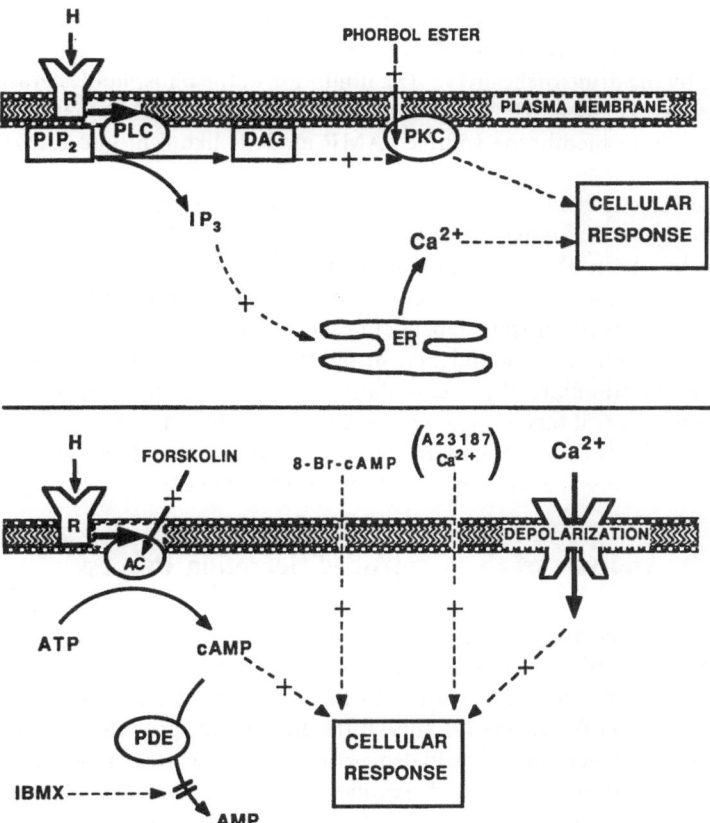

Fig. 2. Common signal transduction mechanisms in mammalian cells. *Top:* Hormone (*H*) binds to membrane receptor (*R*) and causes the activation of phospholipase C (*PLC*). PLC hydrolyzes phosphatidyl inositol [4, 5] bisphosphate (*PIP₂*) to soluble inositol [1, 4, 5] trisphosphate (*IP₃*) and diacylglycerol (*DAG*). IP₃ stimulates the release of Ca^{++} from the endoplasmic reticulum (*ER*) storage, and DAG activates protein kinase C (*PKC*). PKC may also be directly activated by phorbol esters. Ca^{++} and PKC induce other cellular responses.
Bottom: Hormone (*H*) binds to receptor (*R*) and activates adenylate cyclase (*AC*) which converts ATP to cAMP, a potent stimulator of several cellular responses. AC may also be directly activated by forskolin. Phosphodiesterase (*PDE*) hydrolyzes cAMP to AMP; this reaction can be inhibited by isobutylmethylxanthine (IBMX) resulting in increased cytosolic concentrations of cAMP. Nonhydrolyzable, membrane permeable cAMP analogs, like 8BrcAMP, also increase the effective level of cAMP. Depolarization of the plasma membrane opens voltage-sensitive Ca^{++} channels thus allowing calcium to flow from the extracellular medium into the cytosol. Membrane soluble ionophores like A23187 can carry Ca^{++} directly into the cell

now open voltage-sensitive ion channels [1, 7]; and through direct transport of Ca^{++} into the cytosol via membrane-soluble ionophores [9, 10]. An additional mechanism for the increase of cytosolic calcium exists in insulin producing pancreatic islet cells: the metabolism of nutrients such as glucose is coupled to increased flux of calcium through the plasma membrane [10, 11]. Cytosolic cAMP can also increase through several mechanisms: receptor-mediated activation of adenylate cyclase which converts ATP to cAMP [1, 10]; direct activation of adenylate cyclase by the drug forskolin [12, 13]; inhibition of the phosphodiesterase which hydrolyses cAMP to AMP by isobutyl methylxanthine (IBMX) [10, 12]; and exposure of cells to membrane soluble cAMP analogs like 8-bromo-cAMP (BrcAMP) [1, 10].

The precise relationship between cytosolic Ca^{++}, cAMP and induced secretory response remains to be evaluated. In both pancreatic islet and pituitary cells, cAMP elevation has been shown to potentiate Ca^{++} influx through plasma-membrane calcium channels; in islet cells, the mechanism of cAMP-mediated stimulation of secretion is completely dependent on elevation of cytosolic calcium [12, 14]. Some investigators have found that in pituitary gonadotrophs, elevated cytosolic calcium can stimulate the accumulation of cAMP as a precursor to secretion, but it is not clear if this is a universal phenomenon [9]. In general, agents which increase cAMP levels can potentiate the actions of agents which increase calcium levels and vice versa, although exceptions to this rule exist.

1.2 Preliminary Analysis of the Controlled Secretion Process

To elaborate on the engineering advantages of controlled secretion, one must first consider the limitations of conventional cell culture processes. Generally, cell culture product titers are quite low relative to those of typical microbiological processes. There are several reasons for this. First, mammlian growth rates are an order of magnitude smaller than microbial growth rates, so the rate of biocatalyst generation is correspondingly smaller. A second problem is that the biomass density in an animal cell culture is limited either by inherent growth inhibition or by the rate of oxygen transfer. Microbial cultures grow to much higher cell densities in suspension (10^9 cells per ml compared to 10^6–10^7 cells per ml for animal cells) because they are less susceptible to growth inhibition and because vigorous oxygen transfer is more amenable to these systems than to sensitive mammalian cultures. In addition, cell cultures must be provided with adequate volumes of medium to dilute inhibitory metabolic products such as lactate and ammonia to a safe level, so that, of necessity, secreted protein products must be diluted to the same extent.

Low product titers are not insurmountable obstacles to downstream processing. Removal of large quantities of water from the medium is technically straightforward with modern processing equipment. A potentially more serious complication is the fact that cell culture media are contaminated not only by other secreted proteins, but also by exogenous proteins present in growth media. Significant progress has been made towards the formulation of serum-free media, but these need to be supplemented with proteins like albumin, insulin and transferrin. Cell lines amenable to serum-free media may be highly transformed, and as such, may express undesirable traits like inappropriate glycosylation. Conversely, well differentiated cells may grow very poorly in serum-free media. Thus, it is unlikely that the requirement for serum, let alone protein-rich supplements, will be relieved in the near future.

Besides possible technical difficulties, concentration of protein solutions and product purification are costly operations. As the volume of liquid and amount of contaminating proteins increase, the size of downstream processing equipment (e.g. chromatographic columns) increases too, and both capital and operating costs become higher.

To take full advantage of the regulated secretion pathway of properly differentiated cells, protein flux through this pathway should be maximized; this can be achieved by inducing cells periodically to secrete. In other words, cells should be exposed to secretagogue-free medium that promotes protein synthesis and intracellular accumulation, then to secretagogue-supplemented medium to retrieve the protein from storage granules, and the operation should be repeated for as long as it remains efficient. To simplify downstream processing, the secretion medium should be made of as small volume and as simple composition as possible. Such medium cannot sustain cultures for long periods, but cells are certainly able to withstand it for the period of one two hours that is needed to obtain most (if not all) of the induced secretory response.

For production schemes employing regulated secretion to be successful, cells should exhibit high storage capacity and induced secretion rate. Relevant differentiated properties should be retained for the entire duration of a production cycle, expected to be several days. Cells should be able to withstand repeated exposures to secretion and charging medium with minimal loss of their ability for synthesis and induced secretion of active recombinant protein. For a particular cell line, culture conditions should be such as to maximize storage capacity and induced secretion and also help retain the differentiated properties of cells. The bioreactor configuration should permit rapid perfusion of media and minimal diffusional resistances (for nutrients, secretagogue, product).

This article addresses the potential of controlled secretion processes and discusses the above issues using examples from previous work with recombinant mouse AtT-20 [15, 16, 17, 18] and transgenic mouse βTC3 [19] cells. Several topics are of generic importance, since they are common to cell culture operations and secretion of protein products. Problems (observed or expected) and their possible solutions are also discussed. Many of these topics are interrelated, so various issues are discussed from the perspective of each relevant topic.

2 Properties of Cells

2.1 Expression and Secretion of Recombinant Protein

For recombinant proteins to be produced in a regulated secretion scheme, they should be recognized by the sorting and processing apparatus of host cells. It appears that the sorting mechanisms of recombinant differentiated cells recognize only proteins which, in their cells of origin, follow the regulated pathway of secretion [20]. The nature of the sorting "signal sequence" on secretory proteins is presently unknown, but this signal is apparently dominant: a fusion protein derived from a constitutive and a regulated source (truncated vesicular stomatitis virus G protein, molecular weight of $54-56 \times 10^3$, and human growth hormone, molecular weight of 22×10^3, respectively) is secreted via the regulated pathway in AtT-20 cells [20, 21]. In fact, the fusion protein is targeted into the regulated pathway at roughly the same efficiency as parental hGH. These results suggest that regulated secretion may be used for a variety of foreign proteins, as long as the sorting sequence does not become concealed, the packing and function of secretory vesicles is not disrupted, and fusion proteins can be processed to remove fused signal sequences. These requirements are expected to be met with proteins of low to intermediate molecular weight, but it is unknown whether high molecular weight proteins following constitutive secretion can be diverted into the regulated pathway.

To maximize production of a recombinant protein, one should aim at high expression systems, provided that active product is synthesized and secreted efficiently without damage to cells. For complex proteins, there may be a limit to the level of expression which can be accommodated by the processing apparatus of the secretory pathway. An example of this is the apparent retainment in the endoplasmic reticulum of some large or complex macromolecules (i.e. IgG heavy chains or Factor VIII) by a protein known as the immunoglobulin binding protein, or BiP [22, 23, 24]. BiP seems to release proteins to the secretory pathway only if they are correctly folded and associated with the appropriate subunits. Excess protein expression may lead to obstruction of the endoplasmic reticulum by misfolded or mismatched polypeptides which cannot continue through the secretory pathway. In the case of regulated secretion, another possibility is that the capacity of cells to sort and sequester proteins into secretory granules becomes limited by the available amount of specialized regions of the trans Golgi cisternae. Any excess protein would overflow into the constitutive pathway, which is potentially wasteful.

Considerations of cellular economy suggest that cells should contain just enough regulated secretory apparatus to process native secretory proteins, and that the expression of components of this apparatus should be linked to the expression of proteins destined for regulated secretion. Evidence for this phenomenon has been observed in murine pituitary GH3 cells which contain significant quantities of secretory granules and associated proteins only when the synthesis of prolactin or growth hormone is induced [25, 26]. Investigators have shown that in rat pancreatic islets the biosynthesis of secretory granule associated peptides is linked closely to the biosynthesis of insulin, the primary secretory product in these cells [27]. As it has been observed with PC12 cells, and most likely occurs with other cells too, granule membranes and proteins are conserved by recycling from the plasma membrane to the Golgi after an episode of stimulated secretion, which is another indication that their quantities are carefully controlled in secretory cells [28]. Excess synthesis of specialized apparatus (proteins and membranes) would be wasteful of cell resources. Therefore, one would expect that cells could not accommodate a high molar excess of recombinant to native secretory protein if the latter is expressed at normal levels. The recombinant AtT-20 cells with which we work express insulin or human growth hormone (hGH) at approximately the same molar levels as native adrenocorticotropic hormone (ACTH).

In order to increase the efficiency of a production process involving regulated secretion, one must ultimately find a way to repress the native protein while maintaining or enhancing the expression of the regulated secretory apparatus. Initial successes in identifying key molecules involved in regulated secretion and expected future progress in this area, should make the latter of the two objectives feasible within the next few years.

The recombinant protein may be expressed in either a constitutive or a regulated fashion. Biosynthesis is controlled primarily at the level of DNA transcription via a regulated promoter, but additional levels of control are possible through the regulation of the rates of mRNA degradation and translation. For a regulated expression system, the above rates, as well as the rate of intracellular protein degradation, may be controlled, among others, by factors related to the secretory state of cells, such as the presence of secretagogues or inhibitors in the medium or the degree of saturation of the storage capacity of cells.

The native hormone of AtT-20 cells, ACTH, is an example of a regulated promoter system. Transcription of the ACTH gene has been shown to increase following stimulation with secretagogues, and to fall when the cells are exposed to corticosteroids, which act as endocrine feedback inhibitors of ACTH in vivo [10, 29]. Insulin biosynthesis in pancreatic islet cells is controlled at multiple levels in the biosynthetic pathway. During short term glucose stimuli, there is an increase in the specific rate of preproinsulin translation from mRNA templates, and during long term stimulus the specific rate of mRNA synthesis also increases [27].

In the case of a constitutive promoter, protein synthesis is not subject to transcriptional regulation. However, it is possible that induction of cells may lead to elevated rates of total (i.e. intra- plus extracellular) protein accumulation due to the effect of secretagogues on posttranscriptional regulation steps. We have observed that the net rate of synthesis of hGH in recombinant AtT-20 cells may increase by 100% following induction of secretion [17]. It is presently unknown whether this is due to an increased rate of protein synthesis, to a decreased rate of intracellular degradation, or to a combination of these two factors. Intracellular turnover of stored protein through crinophagy (i.e. through fusion of storage granules with lysosomes) can be quite significant, especially in cells with accumulated granules that are not exposed to secretagogue [30, 31].

In general, one would expect the native secretory product of a given cell line to be comprehensively regulated. From the viewpoint of cellular economy, such regulation would be useful in minimizing superfluous biosynthesis. For a cell culture process, in which the principal goal would be to maximize overall productivity, decoupling the regulation of secretion rate from that of biosynthesis would be certainly desirable. For a recombinant product, it is probably a straightforward task to avoid transcriptional down regulation through the use of high-level, constitutive promoters. However, the product may remain subject to regulation by secretion agonists or other factors at additional posttranscriptional steps. Some of these regulatory mechanisms may be amenable to mutagenesis in the non-coding sequences of the recombinant mRNA, but other mechanisms may be inherent to the expression of secretory proteins in the regulated cell line. Future developments in the molecular biology of regulated secretion are expected to offer specific suggestions on the solution of relevant potential problems.

2.2 Growth and Attachment Properties of Cells

A controlled secretion system must incorporate cells immobilized to a macroscpic support in order for efficient perfusion to be feasible. Single cells in suspension cannot easily be retained in a bioreactor during medium changes. This requirement places a number of constraints on the growth and attachment properties of cells. Preferably, cells should be anchorge dependant, in other words, spontaneously self-immobilizing, so that they can be grown as monolayers on flat surfaces or on microcarriers or fibers. Otherwise, special provisions are necessary for retaining the cells in an immobilized phase. These include encapsulation, or entrapment in a porous support or in the extracapillary space of a hollow fiber reactor.

Encapsulated or entrapped systems may provide a certain benefit in that they protect cells from hydrodynamic forces in the reactor flow field, but they are inherently more complex than monolayer cultures, and they have an additional disadvantage specifically relevant to controlled secretion. Diffusional limitations in a dense, immobilized cell mass may prevent the efficient penetration of secretagogues and the removal of macromolecules during the protein recovery cycle. For example, the time constant for diffusional equilibrium of a 10 kilodalton solute in a 200 μm layer of biomass is approximately one hour assuming a diffusion coefficient of 10^{-7} cm^2 s^{-1} [32]. Thus, even if nutrient transport is sufficient to maintain the cell culture at steady-state, transport of a secretagogue and of large polypeptides may be too slow for optimal recovery during the transient period of induced secretion.

There are also biological considerations for using an anchorage dependent cell line. All of the primary endocrine and exocrine cells are anchorage dependent and require appropriate cell-substrate and cell-cell contacts for correct morphology and function. These constraints normally apply to the established cell lines initially derived from primary cells. As cells become more transformed or tumor-like, they may lose anchorage dependence, but they may also lose differentiated functions such as specific post-translational processing or regulated secretion. We have observed that AtT-20 cells may lose specialized properties within the lifetime of a single culture: prolonged exposure to growth medium results in an outgrowth of attached, spread cells by apparently more transformed round cells in foci which exhibit inferior regulated secretory response [17]. Loss of properties of insulin-producing AtT-20 cells also occurs with multiple passages in culture (reference and in Fig. 6). Similarly, the same cells grown as spherical aggregates (spheroids) have an inferior regulated response compared to monolayer cultures [unpublished data]. Other investigators have noted loss of expression and regulated function with time in cultures of various endocrine cell lines (GH3, HIT, βTC3) [19, 33, 34]. Surprisingly, this process may even be reversible: neuroblastoma cells can be induced to express regulated secretory apparatus de novo when exposed to differentiating agents [35]. Thus, maintenance of one differentiated property (regulated secretion) may be inherently linked to the maintenance of another (anchorage dependence).

Ideally, one would like the cells to grow quickly and, under appropriate conditions, remain as stable, confluent monolayers for prolonged periods of time. Tumor cells (e.g. insulin-producing AtT-20) have good growth properties (doubling time of 28 hours), but retaining them as stable layers of differentiated cells is difficult. Layers may be unstable and detach (e.g. insulin-producing AtT-20 cells), or they remain attached, but undifferentiated cells in foci (which lack the property of interest) may outgrow attached, differentiated cells (e.g. hGH-producing AtT-20 cells) [17, 18]. We have been able to prevent detachment of AtT-20 insulin cells by pretreating the culture substrate with polylysine, but the outgrowth of foci has proved intractable.

The maintenance of proper cell structure and function may require a more fundamental understanding of the mechanisms relating cellular differentiation to interactions between cells and their environment. For example, the supplementation of certain extracellular matrix proteins such as laminin has been shown to induce and maintain the optimal cellular morphology and regulated functions of GH3 and AtT-20 cells in culture [36, 37]. A systematic investigation of cell maintenance in media that do not promote excess growth, and of the effects of protease inhibitors, extracellular, exogenous matrices, and hormonal signals may yield a combination of such factors that allows monolayers of cell to remain stable for prolonged periods of time [38].

2.3 Secretion Kinetics

The relevant properties of secretion kinetics must be studied in the context of at least three phases of the controlled secretion cycling process: basal secretion of fully charged cells (i.e. of cells not recently exposed to secretagogue), stimulated secretion of charged cells, and recharging of depleted cells (i.e. of cells exposed to secretagogue for one or more hours). In the absence of secretagogue stimulation, basal secretion kinetics depend upon the degree of regulation of the rates of gene transcription and translation and protein degradation and transport. If the promoter to the gene of interest is constitutive, cells may synthesize protein at a roughly constant rate during all phases of the secretion cycle. Fully charged cells could process some of this protein through the regulated pathway as new capacity becomes available through cell growth and division or turnover of mature granules, but most of the excess protein will be degraded or secreted constitutively (unmodified by the enzymes of the regulated pathway).

The time courses of secreted and intracellular hGH (experiencing little, if any, transcriptional regulation) in cultures of recombinant AtT-20 cells are shown in Fig. 3. Basal secretion was observed in serum-free Dulbecco's modified Eagle's medium (DMEM) and the induced response in serum-free DMEM supplemented with 5 mM BrcAMP. Basal secretion occurred at a roughly constant rate of 0.32 ng h^{-1} per 10^5 cells, and intracellular hGH in uninduced cells remained

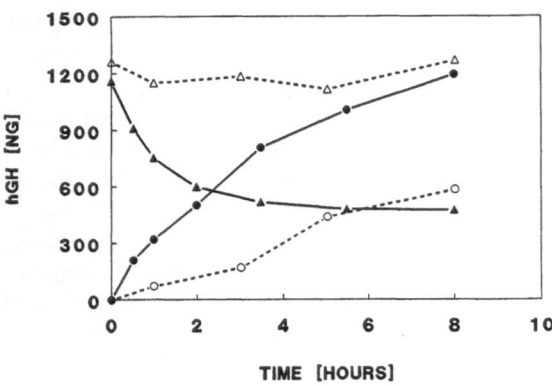

Fig. 3. (from 17). Basal and induced secretion of recombinant hGH from AtT-20 cells. Cells in 25 cm² T flasks were exposed until time zero to complete growth medium [Dulbecco's modified Eagle's medium (DMEM) with 4.5 g l^{-1} glucose, supplemented with 10% fetal bovine serum] and at that time were switched to serum-free DMEM with or without 5 mM 8BrcAMP (3 ml in each culture). At various times, media were collected and cells extracted with a detergent-containing buffer. hGH in media and extracts was assayed by radio-immunoassay using a commercially available kit. The average number of cells in each culture was 2.37 × 10⁷.
----○---- hGH secreted in DMEM;
——●—— hGH secreted in DMEM with BrcAMP (DMEM-S).
----△---- intracellular hGH in cells exposed to DMEM;
——▲—— intracellular hGH in cells exposed to DMEM-S

constant at 5 ng per 10⁵ cells. Cells exposed to BrcAMP secreted at an initial rate of 1.7 ng h^{-1} per 10⁵ cells, which eventually decreased to that of basal secretion; intracellular hGH declined sharply to a value of 2 ng per 10⁵ cells. At the end of the experiment, induced cells had secreted an amount of 2.5 ng per 10⁵ cells more than controls; an amount of 0.5 ng per 10⁵ cells apparently underwent degradation in the secretagogue-supplemented culture. Although it took 4–5 h for induced cells to return to the basal rate, most of the stored incracellular hGH was secreted during the first two hours of exposure to secretagogue. For cells exhibiting such kinetics, an efficient process would limit the period of induced secretion so that recovery and recharging of the storage capacity would commence promptly once the greatest part of stored protein had been recovered. Presumably, this time period (1 to 2 h) is short enough for cells to withstand exposure to secretion medium without any major intrinsic damage from the lack of serum supplements or from stimulation by secretagogues.

If the level of mRNA transcribed from the gene of interest is regulated (e.g. βTC3 cells), protein synthesis may be very low in unstimulated cells. The protein could have any of the fates outlined above for a constitutive promoter, but will be more easily accommodated by the basal turnover of mature storage granules. In terms of the basal secretion kinetics there is no clear advantage a priori in having a regulated or constitutive promoter: the period during which cells are maintained at their fully charged state should be relatively short, so the "wasted" constitutive secretion should be minimal in either case. However, if the rates of protein synthesis and secretion are too tightly coupled in a cell line with regulated gene expression, the recharging process may be inefficient (see Sect. 2.5).

The kinetics of stimulated secretion are very important. To obtain the full benefits of regulated secretion, the storage pool should be drained quickly and completely following stimulation with a secretagogue. The secretion kinetics depend upon the inherent properties of cells and the nature of the secretagogue. The signal transduction pathways are subject to complex phenomena including inhibition and down regulation, which act to limit the strength or duration of a response; and potentiation and synergism, which can strengthen and extend a response. A secretagogue or secretagogue combination should limit the former and exploit the latter phenomena to achieve the desired secretion kinetics. The secretagogue treatment should also avoid residual effects upon the cells which might impair the process of recharging of the cells after an episode of stimulated secretion (see Sects. 2.6 and 3.1).

2.4 Storage Capacity of Cells

Whether protein transcription is constitutive or regulated, there should be good coordination between the rate of protein synthesis and the storage capacity of cells. Here, storage capacity is defined as the portion of the total intracellular pool of secretory protein which can be released rapidly during an episode of induced secretion. An additional portion of "unavailable" protein may be recovered in cell extracts but not immediately through stimulated exocytosis, presumably because it is in the endoplasmic reticulum, the Golgi apparatus, or in transit through other parts of the secretory pathway. In the context of a cycling controlled secretion process, cells should possess a storage capacity corresponding to the net amount of protein synthesized over a period of at least ten hours; this amount should be recoverable in a relatively short period of stimulated secretion (1 to 2 h). Such properties would ensure that synthesis/recovery cycles are not too frequent.

In recombinant AtT-20 cells, the coordination between the storage capacity and the synthesis rate of recombinant protein is just within the limits of the above-mentioned criteria. Recombinant cells store roughly 30–35 microunits [μU] of insulin or 3 ng of hGH per 10^5 cells, while the synthesis rates are 3–5 μU h^{-1} per 10^5 cells and 0.3–0.8 ng h^{-1} per 10^5 cells for insulin and hGH, respectively. The cells thus store an amount corresponding to the protein synthesized over a period of 4–10 h; and, since 1 to 2 h are required for recovery of most of the stored protein, the ratio of dynamic compression (recovery rate: synthesis rate) is about 5:1. In practice, the frequency of cycling is much lower than that suggested by the foregoing ratio due to secretion of synthesized protein by cells being recharged, which prolongs considerably the duration of recharging priods (see Sect. 2.5).

A much more promising alternative is offered by the cell line βTC3 which expresses native insulin on a regulated promoter. In this case, the rate of insulin synthesis varies with the state of the signal transduction pathways and of the metabolism of cells. For example, the synthesis rate increases from 10 to 70 $\mu U\ h^{-1}$ per 10^5 cells when the concentration of glucose in the medium is changed from 0 to 1 mM. Concurrently, the rate of insulin secretion increases transiently from 10 to 200 $\mu U\ h^{-1}$ per 10^5 cells in response to this glucose stimulus, which results in depletion within a few hours of a large fraction of the available storage capacity of 600–800 μU per 10^5 cells. The secretory response can be further potentiated with additional secretagogoues to a rate greater than 400 $\mu U\ h^{-1}$ per 10^5 cells without an increase in the rate of synthesis. Since cells cannot be maintained for prolonged periods in glucose-free medium, a realistic basis for the evaluation of storage capacity and dynamic compression is the rate of insulin synthesis in medium with 1 mM glucose. In these terms, βTC3 cells have a 10 h storage capacity and a maximum ratio of dynamic compression (when secretion is induced by glucose and other secretagogues) of about 6 : 1. It should be noted that, like AtT-20 cells, βTC3 cells land tend to secrete significant amounts of protein during recharging. However, we have been able to correct this problem by manipulating the signal transduction pathways (see Sect. 2.5). For this reason, and because βTC3 combines a relatively high specific productivity with favorable secretory dynamics, it is a much more promising system for production operations than AtT-20 cells.

2.5 Charging Kinetics

Ultimately, the success of the proposed process depends upon the efficiency of accumulation of storage granules following an episode of induced secretion. The faster the cells are recharged to the original capacity, the better. Fastest recharging would be that in which all synthesized protein is directed into the storage compartment. If small, the amount of protein secreted during recharging can be ignored, so in production operations all protein is retrieved from induced secretion media. This alleviates the need for designing two separation processes. In the case of AtT-20 cells expressing hGH, recharging occurs in 20 h, which is twice the expected period with the basal rate of hGH synthesis. The cause of secretion during recharging is unclear. Three possibilities, alone or in combination with one another, exist: a) There is an after-effect of the secretagogue (specifically, BrcAMP) on cells, so that secretory vesicles fuse with the plasma membrane at a high rate for several hours after the secretagogue has been removed from the medium. This is an unlikely possibility, but it cannot be ruled out on the basis of presently available data. b) Missorting of hGH occurs at the trans-Golgi, so a significant amount of hormone enters the transport vesicles of the constitutive pathway. Mathematical models of intracellular protein trafficking suggest that this is indeed a likely possibility in cells switched from secretion to growth medium [39]. c) The sorting efficiency at the trans Golgi is high, but only a fraction of clathrin-coated vesicles mature to storage granules; the rest fuse with the plasma membrane in an essentially constitutive fashion. Mathematical models in which the rate of maturation of vesicles is negatively correlated to the level of saturation of the storage compartment are in agreement with the experimentally observed difference between total and intracellular hGH accumulations [39].

If BrcAMP has a long-term effect on cells, other secretagogues, alone or in combination, may not, and use of those may reduce the recharging time. If missorting occurs, lowering the expression of endogeneous, regulated secretory proteins or increasing the expression of sorting proteins by recombinant techniques (not possible yet) may again help cells recharge faster. If mechanism c) above is the controlling one, it is not clear what modifications, if any, would reduce secretion during recharging.

With the βTC3 system, we found that secretion during recharging can be moderated by external signals down to a minimal, basal level. This suggests that the undesired secretion occurs primarily through the regulated pathway. The precise relative amounts of insulin and proinsulin secreted during recharging are presently unknown, but the foregoing results suggest that there is a significant amount of mature insulin released into the medium. Mature insulin secretion during recharging is conceivably wasteful because the peptide, which would otherwise be recovered during the induction period, is retrieved in relatively dilute form in serum-containing medium.

The problem with the βTC3 system is that protein secretion is tightly linked to the rate of protein synthesis. Consequently, when protein synthesis is stimulated to reasonable rates (in high glucose), exocytosis is stimulated to a similar extent, resulting in inefficient recharging. It is possible to inhibit secretion by denying these cells glucose or by providing a specific inhibitor of secretion. An example of the latter is adenosine which binds to inhibitory receptors on the beta cell membrane. This binding event is transduced to effect a reduction of cytoplasmic cyclic AMP levels with the consequent inhibition of insulin secretion [40]. We attempted to improve the recharging kinetics by using low glucose (0.1 mM), by substituting glucose with slowly metabolized galactose, or by administering adenosine in the presence of high glucose. In all three cases we found that the rate of insulin secretion was reduced, as expected, but the rate of synthesis was limited proportionally so that the net rate of recharging actually declined relative to the high glucose control (see Fig. 4).

While it is not known how signals from glucose metabolism or other stimuli are transduced to control the levels of insulin synthesis in beta cells, investigators have clearly established that intracellular free calcium is the ultimate effector of stimulated protein secretion in pancreatic beta cells [12]. Studies with rat islets have shown that insulin secretion can be attenuated, without affecting insulin biosynthesis, by chelating extracellular calcium [27]. It is reasonable to suggest that a direct manipulation of the intracellular calcium level would bypass "upstream" signal-transduction events related to regulation of protein synthesis. This would allow decoupling of insulin exocytosis from synthesis, so that we could maintain a low secretion level concomitant with a high synthesis rate.

We have verified that the synthesis-secretion coupling can be broken in βTC3 cells by direct manipulation of the calcium fluxes across the cell membrane [41, 42]. This may be achieved by reducing the calcium concentration in the medium, closing the voltage-sensitive calcium channels by hyperpolarizing the cells with low potassium, or blocking the channels directly with a calcium channel blocker [12, 43]. Use of media that incorporate one or more of these treatments (calcium-

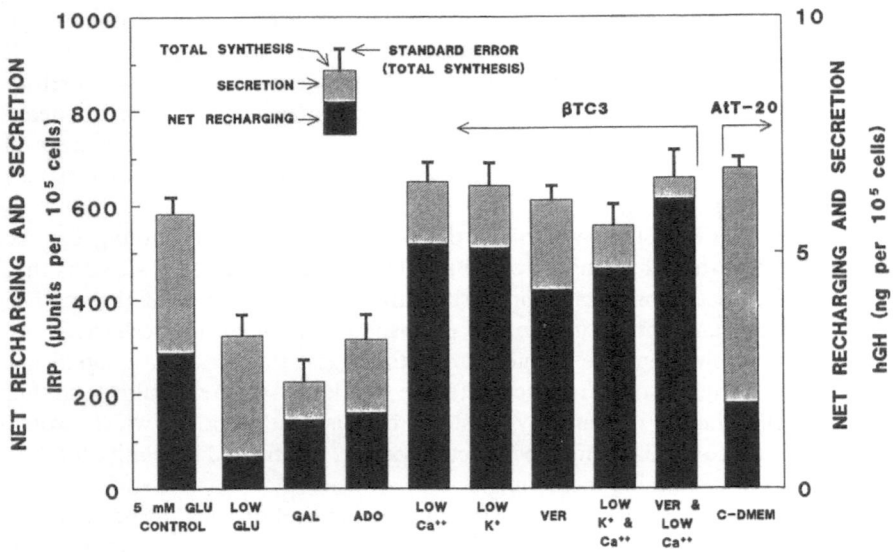

Fig. 4. (from 41, 42, 17). Net amounts of intracellular peptide accumulation (recharging) and secretion for βTC3 and hGH-producing AtT-20 cells exposed to various recharging media following induction of secretion.

βTC3 cells in T flasks were exposed to DMEM with 5 mM glucose and 1 mM IBMX (I-DMEM) for 1.5 hours to deplete intracellular insulin stores. The flasks were then switched to recharging media, and samples of supernatants and of detergent extracted cells were taken after eight hours (except for LOW GLU- see below). The amounts of insulin-related peptides (IRP) in supernatants and extracts were determined using a commercially available radioimmunoassay kit, and the net amounts of recharging and secretion were estimated.

Recharging experiments compared the various treatments to a control medium with 5 mM glucose (5 mM GLU). In one experiment using DMEM with 0.1 mM Glucose (LOW GLU) the medium was replaced every 3 h, up to 9 h to prevent glucose depletion. The data at 9 h were linearly interpolated to 8 h for comparison with the rest of results. For the other experiments. DMEM substitutes were prepared that contained similar levels of buffers, salts, and amino acids, but with $CaCl_2$ or KCl adjusted from 1.8 mM and 4.5 mM, respectively, to other proper concentrations. With the exception of the GAL treatment, these media contained 5 mM glucose with deviations from normal DMEM (CONTROL) as follows: 5 mM Galactose (GAL), 0.1 mM Adenosine (ADO), 0.5 mM $CaCl_2$ (LOW Ca^{++}), 1 mM KCL (LOW K^+), 20 μM verapamil (VER), 0.5 mM $CaCl_2$ and 1 mM KCl (LOW K^+ & Ca^{++}), or 20 μM verapamil and 0.5 mM $CaCl_2$ (VER & LOW Ca^{++}).

All of the calcium depleting treatments (verapamil, low calcium, or low potassium) successfully reduced secretion during recharging relative to the control while maintaining high levels of total insulin synthesis. The metabolism-inhibiting treatments (galactose or low glucose) and the cAMP inhibitor adenosine depressed secretion but failed to maintain an adequate level of synthesis.

hGH-producing AtT-20 cells in T flasks were exposed for 3.25 h to DMEM with 4.5 g l^{-1} (25 mM) glucose and supplemented with 5 mM 8BrcAMP, and were then switched to secretagogue-free growth medium (DMEM with 10% fetal bovine serum). The amounts of hGH in spent growth medium and detergent-prepared cell extracts were determined with a commercially available radioimmunoassay kit. The displayed values were obtained by linear interpolation to 8 h of results obtained at 6.5 and 9.9 h after the medium change attenuating media) dramatically reduces the rate of insulin secretion from βTC3 cells in the presence of high glucose, but it has little or no effect on the rate of insulin synthesis. The net effect is a significant improvement in recharging efficiency (Fig. 4) [41, 42]. We have not investigated the effect of similar approaches to secretion during recharging of AtT-20 cells.

2.6 Tolerance to Repeated Cycling

There is no reason to expect a priori that cells will be damaged in cycling operations by repeated exposures to secretagogues: cells of endocrine and exocrine glands work this way in vivo. Nevertheless, it is conceivable that artificial secretagogue combinations could stress cells to a greater extent than do natural hormonal cycles in vivo. For example, insulin-producing AtT-20 cells which were cycled repeatedly between secretion medium with BrcAMP and growth medium responded well to BrcAMP during the first three episodes of induced secretion, but started to detach much sooner than cells in control cultures never exposed to secretagogue. Consequently, cycling failed prematurely due to the removal of detached cells during medium changes [18].

It is possible that early detachment is due to an enhancement of total metabolic activity by exposure to BrcAMP. We have noted that cells exposed to this secretagogue tend to secrete excess lactate into the medium, and detachment is more likely to occur in acidic spent medium due to activation of proteases. Another possibility is that BrAMP has more specific effects on cell morphology and function. Investigators have reported that chronic exposure of AtT-20 cells to dibutyryl cAMP can lead to cell enlargement and impairment of regulated secretion [44]. It may be possible to prevent detachment by growing cells on proper extracellular matrices and supplying medium with protease inhibitors.

As with their insulin-producing counterparts, cultures of hGH-producing cells worked satisfactorily in cycling experiments for the first 3–5 cycles. Longer operations failed, but for different reasons. Namely, attached, differentiated cells were outgrown by round cells in foci that appeared to exhibit inferior, if any, response to secretagogues (Fig. 5a, b) [17]. Cells in foci do revert to the attached, spread morphology after proper trypsinization and seeding at low density. Outgrowth of foci may thus be delayed by thoroughly removing foci aggregates from the inoculum to a culture (through repeated pipetting and controlled trypsinization) or even totally prevented by recharging cells in a medium that supports cell maintenance and intracellular protein accumulation, but no cell growth.

The βTC3 cell line does not exhibit instabilities in cell morphology or attachment and can be cultured to high densities on tissue culture plastic without loss of regulated secretion activity. In a preliminary experiment we used calcium depleting medium (see Sect. 2.5) to recharge cells between episodes of induced secretion with 1 mM IBMX. Cultures were cycled seven times with only a slight loss of specific productivity over the period of seven days. This was adequately compensated by an increase in cell density, so the overall rates and amounts of secretion remained at roughly the same levels throughout the experiment (Fig. 5c, d) [41, 42].

We have noted several unexpected effects of repeated cycling between secretion and recharging media. Transient exposure of the cells to IBMX in the secretion medium causes an enhanced rate of glucose metabolism which persists for several hours after the IBMX is removed. Care must be taken to provide ample perfusion of medium to prevent the accumulation of inhibitory concentrations of lactate in cultures being recharged (see Sect. 3).

We have also noted that exposure of βTC3 cultures to the special calcium-reducing, recharging medium depresses the growth rate slightly with respect to controls which are recharged in normal growth medium. The reduced growth rate may be a consequence of low levels of intracellular calcium which is an ubiquitous messenger for regulation of many cellular functions besides regulated secretion. Alternatively, growth may be stunted due to the hyperpolarization of the cell membrane which could alter the trans-membrane transport of ions besides calcium. In any case, this phenomenon is not linked to deleterious effects on cell function or viability, and may even be desirable if the cycling operation can be extended under conditions of slower growth.

Fig. 5a–d. (from 41, 42, 17). Cycling of hGH-producing AtT-20 cells and of βTC3 cells ▶ between growth and secretion media. Cells in T flasks were exposed for periods of between 21 and 24 h to complete growth medium and were then switched for 1.5 to 2 h to secretion medium. Supernatants were sampled at the end of each incubation period, and the amounts of hGH (AtT-20 cells) or insulin-related peptides (IRP, βTC3 cells) were determined by radio-immunoassay using commercially available kits. Figures **5a** and **5c** show the total amounts of hGH and IRP secreted during each phase of the cycle (growth and induced secretion). The bars are positioned roughly at the end or the each corresponding phase. The growth and recharging phases are indicated by the open bars and induced secretion phases are indicated by the filled bars. Figures **5b** and **5d** show the average rates of secretion of hGG or IRP during each phase of the cycle. Here, the widths of the bars represent the duration of the phase so that the areas of the bars correspond to the relative amounts of protein secreted during each phase. Growth phase data are indicated by hatched bars, while the bars for induced secretion are filled.
a, b: AtT-20 cells expressing hGH were cycled between DMEM with 4.5 g l^{-1} glucose supplemented with 10% fetal bovine serum (C-DMEM) and serum-free DMEM with 5 mM 8BrcAMP (DMEM-S) for the 2 h induction period. The amount of hGH retrieved during the induced secretion periods was only 13.5% of total secreted hGH, and the ratio of induced to non-induced secretion rates was poor throughout the experiment.
c, d: βTC3 cells expressing insulin were recharged in a DMEM substitute (R-DMEM) containing reduced concentrations of calcium and potassium (0.5 mM and 1 mM respectively) and supplemented with horse serum and fetal bovine serum at 2.5% each. Cultures were induced for 1.5 h with DMEM containing 5 mM glucose and 1 mM IBMX (I-DMEM). Note that, in contrast to the AtT-20 cells, most of the product protein (% of total) was recovered in the induction medium

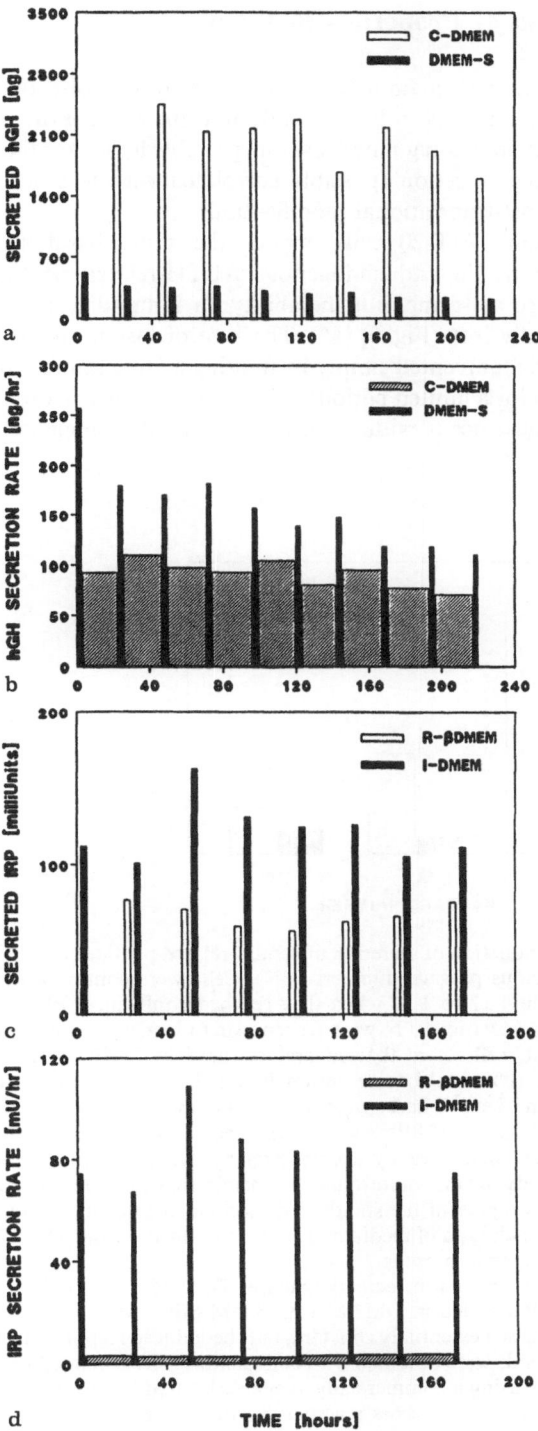

2.7 Retainment of Properties In Vitro

A general problem with mammalian cells is that of loss of properties as passage number increases, i.e. as cells are cultured in vitro. For cells secreting a recombinant protein in a regulated fashion, possible losses are those of expression, ability for induced secretion (possibly correlated with loss of storage capacity) and ability for post-translational modifications.

Insulin-producing AtT-20 cells exhibit the same basal secretion rates of recombinant proinsulin and endogenous ACTH-related peptides after several months of in vitro culturing, but the ability for induced secretion of insulin and ACTH is gradually lost (Fig. 6) [45]. The rate of loss is small, however, so the problem can be circumvented simply by freezing a large number of cells and using thawed cells only for a limited period of time (e.g. few weeks). Loss of differentiated properties may also occur within the time frame of a single culture: if cells are

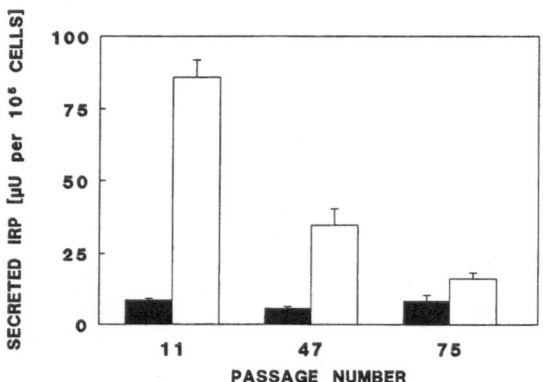

Fig. 6. (from 45). Induction of secretion of insulin-related peptides (IRP) from recombinant AtT-20 cells of various passage numbers (PN). Cells were continuously propagated in T flasks and were split 1 : 2 to 1 : 6 when they reached confluency. PN increased by one at each splitting. Cells at various PN were frozen using identical procedures.
For the experiment, cells were thawed and cultured in DMEM with $4.5\,\text{g}\,\text{l}^{-1}$ glucose supplemented with 10% fetal bovine serum (C-DMEM) for a sufficiently long period to fully recover. Cells in $25\,\text{cm}^2$ T flasks were switched at time zero from C-DMEM to serum-free DMEM with or without 5 mM 8BrcAMP. These media were sampled at 2 h, and the amounts of secreted IRP were measured by a commercially available radioimmunoassay kit that employed an antibody against mature human insulin that has a lower cross-reactivity with proinsulin. The bars represent the standard deviations of three independent evaluations at each PN and with each type of medium. The increase in PN from 11 to 75 corresponds to a time period of roughly 5 months.
■: amounts of IRP secreted in secretagogue-free DMEM
□: amounts of IRP secreted in DMEM with 5 mM 8BrcAMP.
Basal secretion remained essentially constant, but the induced response declined significantly with PN. Similar trends were obtained with endogenous ACTH-related peptides, which were also assayed by RIA using a commercially available kit that employed an antiserum against mature ACTH that had lower cross-reactivities with ACTH precursors

exposed for prolonged periods to growth-promoting medium, cells with more expressed tumor phenotypes may outgrow slow-growing differentiated cells, so that the overall culture properties are gradually lost. Maintaining a stable layer of spread tumor cells may be an intricate job requiring a combination of proper medium, hormonal signals from the liquid phase and additional signals from the solid extracellular matrix underlying the cells [38].

3 Culture Conditions

Protein synthesis, maturation and secretion are energy-requiring processes, so the medium bathing the cells should contain adequate amounts of nutrients, especially the energy-providing glucose and glutamine. Insulin-producing AtT-20 cells exposed to medium without glucose and glutamine exhibited inferior secretory responses compared to cells exposed to the same media supplemented with these compounds [45, 46]. In these experiments, secretion was induced by BrcAMP or by a combination of IBMX, membrane depolarization, and elevated Ca^{++} concentration in the medium. In the same context, adequate amounts of oxygen should be provided to cultures of secreting cells, particularly when cells are induced with a secretagogue. Recent experimental findings show that insulin-producing AtT-20 cells exposed to DMEM with 5 mM BrcAMP consume oxygen at a rate which is significantly higher than that of cells exposed to secretagogue-free DMEM [S. Park, personal communication]. We have observed a 2-fold stimulation of the glucose metabolism rate of βTC3 cells exposed to medium with IBMX.

Such high metabolic rates may be accommodated in monolayer cultures which are not subject to diffusional limitations. However, cells grown as multilayered aggregates (in spheroids, gel capsules, hollow fibers, ceramic matrices, or porous microcarriers) are more susceptible to the depletion of oxygen and other nutrients. For this reason, processes using these types of systems would have to be designed carefully to provide for the efficient transport of nutrients as well as product during the transient period of induced secretion (see Sect. 4).

One must also consider the potential benefits of auxiliary factors in the growth and secretion media. For example, differentiating factors or extracellular matrix components may promote or maintain differentiation of the cell culture for improved regulated function [35, 36, 47]. Other agents may inhibit the undesirable exocytosis of secretory granules during recharging, which is especially important for βTC3 cells. The regulated expression system for insulin of these cells may even offer the possibility of selectively enhancing the rate of insulin synthesis by manipulating the signal-transduction pathways with hormones or other agents. This would provide a means to increase the productivity without resorting to gene amplification. The high product titer in secretion medium raises also the issue of the stability of secreted protein, especially if the medium is saved for multiple episodes of induced secretion. We have observed that substantial loss of insulin titer may occur in spent secretion medium from AtT-20 cultures, so addition of protease inhibitors and other stabilizers to this medium should be considered.

3.1 Secretagogues

The most obvious reason for choosing a secretagogue is that it induces a strong secretory response, but there are a number of other properties which are relevant to a controlled secretion process. For example, secretagogue cost may be an important factor. The cyclic nucleotide analog BrcAMP is a potent secretagogue for AtT-20 cells. It mimics the effects of intracellular synthesis of cAMP while resisting deactivation by the endogenous phosphodiesterase responsible for rapid downregulation of cAMP. However, BrcAMP is prohibitively expensive for larger than lab-scale operations, so we have developed alternative secretagogues which take advantage of the synergism between the effects of calcium and cAMP. With insulin-producing AtT-20 cells, we have found that inhibition of the cAMP phosphodiesterase with methyl xanthines, specifically IBMX, can potentiate the

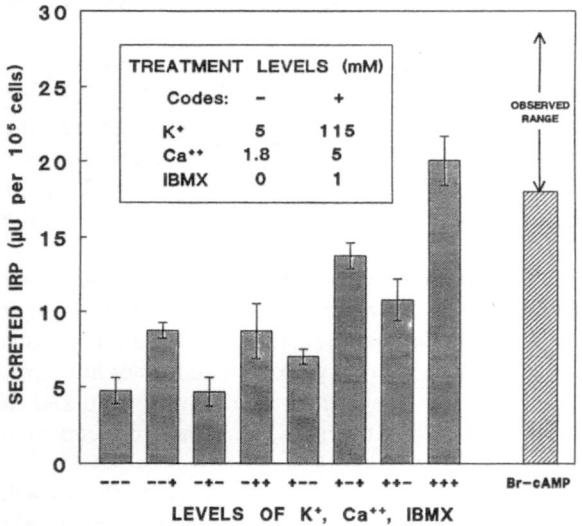

Fig. 7. (from 42, 48). Synergism of secretion treatments. AtT-20 cells expressing human proinsulin were grown in T flasks using DMEM with 4.5 g l^{-1} glucose supplemented with 10% fetal bovine serum. For the experiment the cultures were washed and switched to serum free DMEM substitutes and the supernatants were sampled after two hours. Insulin-related peptides (IRP) were measured by radioimmunoassay using a commercially available kit. The secretion media were formulated with various concentration of IBMX, calcium and potassium but were otherwise similar to DMEM in terms of nutrients, salts, and amino acids. IBMX was absent (−) or present at 1 mM (+), calcium chloride was at the normal DMEM level of 1.8 mM (−) or elevated at 5 mM (+), and potassium chloride was at the normal level of 4.5 mM (−), or at the depolarizing level of 115 mM (+). The concentration of sodium chloride was reduced to maintain a constant ionic strength in the media with elevated calcium or potassium.
Used individually, each of the three modifications to DMEM induced little or no excess secretion relative to the (−−−) control. The combined effect of IBMX with elevated calcium and potassium compares well with the effect of 5 mM 8BrcAMP

effect of depolarization of the plasma membrane to yield an excellent secretory response (Fig. 7) [48]. Depolarization, which is accomplished by substituting K^+ for Na^+ in the medium, results in elevated cytosolic free Ca^{++} concentrations by opening voltage-gated calcium channels localized in the membrane and thus allowing Ca^{++} to flow from the medium into the cytosol [1]. This effect is potentiated by increasing the concentration of Ca^{++} in the medium.

A second issue that needs to be considered is that of a possible after-effect of the secretagogue on cells. As mentioned above, secretion during recharging of hGH-producing AtT-20 cells may be the result of such an after-effect of BrcAMP. An "ideal" secretagogue should have minimal after-effects, so that cells can start recovering and accumulating protein immediately after one episode of induced secretion. In the case of a regulated promoter, some residual effects may be beneficial: if the protein synthesis rate is significantly elevated after exposure to secretagogues the recharging period may be shorter, provided that the secretion rate is not proportionally elevated.

We have found that the onset of net protein accumulation in βTC3 cells in normal growth medium is delayed by up 3 hours following exposure to medium with IBMX. This is due to a high level of residual secretion, which is probably associated with the elevated glucose uptake rate occurring over a similar period. Use of a calcium-reducing medium can minimize this residual secretion during recharging.

4 Engineering Issues

Issues of bioreactor design, operation and control become more important as cultures are scaled up. Bioreactors should allow for rapid perfusion of media (so that cycling operational schemes can be easily realized) and diffusional resistance should be minimal. Such resistances are indeed minimal across cell monolayers, but in case of spheroids (see Sect. 2.2) they can be significant. To avoid cores that are anoxic or that do not receive secretagogue during an induction period, the size of spheroids should be kept (mechanically or otherwise) below a certain limit; calculations show this limit to be of the order of 400 micrometers [49]. Even if the transport of low molecular weight species is not limiting, the impeded recovery of larger product proteins from a dense cell matrix could be a concern as mentiond previously (see Sect. 2.2). Diffusional resistances may also become important if cells are grown inside porous microcarriers, gel beads, or ceramic matrices. Despite these potential problems, controlled secretion is not necessarily incompatible with high density cell culture: if the reactor is designed to provide significant convection through the cell mass, transport rates are improved dramatically [32, 50].

One must also consider the operational mode of a bioreactor with respect to controlled secretion. A stirred tank containing cells attached to non-porous or porous microcarriers is an excellent scaleable system but may present problems during the rapid perfusions that are needed in cycling operations. Simple gravitational settling may be too slow even for microcarrier cultures when frequent perfusions are necessary. Other retainment devices based on gravity, like inclined sedimentors, may be sufficient for the continuous supply of nutrients and removal of waste products, but again do not appear appropriate for frequent, rapid alternations between growth and secretion media. Devices based on size exclusion, such as spinning filters, seem more appropriate, provided that they do not become easily clogged by suspended cells and cell debris. Hollow fiber reactors allow for rapid perfusion of media in the lumen of the tubes, but may be unsuitable for controlled secretion processes. Besides the significant diffusional resistances encountered in these systems, which slow down the exchange of compounds in the extracapillary space, the method of growth of cells as aggregates rather than monolayers implies an inferior induced secretory response (a phenomenon we have observed in both AtT-20 and βTC3 cells).

More attractive alternatives are those of packed or fluidized beds of cell carriers. Fluidized beds assure relatively good transport and mixing within the bed and allow for rapid and efficient switching between growth and secretion media, but flow rates need to be carefully controlled to avoid washout or bed settling. This may reduce one's flexibility to operate the reactor optimally during the very distinct growth and secretion modes. Packed beds can tolerate a wider range of flow rates but are subject to problems of bed compaction, clogging and channeling. The possibility of such complications should be carefully evaluated to avoid compromising the health of the culture and the efficiency of protein recovery. Experiments in which AtT-20 cells are grown in rigid, porous weighted carriers in a fixed bed configuration have demonstrated that convective transport through the pores is sufficient to overcome diffusional limitations, that the cells grow well in this system and that they retain a regulated secretory response which is comparable to that observed in monolayer cultures [Seujeung Park, personal communication, [50]].

In most fixed-bed bioreactors the medium is oxygenated in a separate chamber and recycled through the bed. This is because the specific rate of oxygen depletion exceeds the depletion rate for other medium constituents by orders of magnitude. Typically, the medium residence time for a single pass through the reactor bed is on the order of minutes while the residence time for medium perfused through the entire system is on the order of hours. This situation precludes the optimal utilization of secretion medium in a controlled secretion process because the time required for quantitative recovery of stored proteins is on the order of one to two hours. Thus, a batch of secretion medium must be reoxygenated and recycled during one episode of induction. The medium itself is only partially depleted in nutrients after a single induction but could be reused over several induction cycles (as long as the secretagogues and nutrients retain the potential to support stimulated secretion).

An alternative fixed-bed design which we have explored incorporates in-situ oxygenation of the bed. Specifically, a porous ceramic monolith is fitted with oxygen-permeable silicone tubes in each of the macroscopic flow channels so that oxygen is provided to the cells by local diffusion throughout the entire length of the reactor [51]. This design obviates rapid recirculation of the medium through the reactor and actually allows the system to be operated as a single-pass or "plug-flow" bioreactor. We tested this design with βTC3 cells and found that we could manipulate the residence times such that conditions during growth, recharging, washing, and induced secretion were indpendently optimized. In particular, we could introduce secretion medium with residence times of between one and three hours such that stored protein was quantitatively recovered concomitant with efficient utilization of the medium in a single pass through the reactor.

5 Discussion and Conclusions

We have presented an evaluation of various issues involved in the development and optimization of novel controlled secretion processes aimed at simplifying downstream processing and ensuring biological activity of secreted product. These processes can offer significant advantages relative to traditional operations employing constitutive secretion, provided that certain criteria regarding the properties of cells, the culture conditions and the bioreactor configuration are met.

Cells exhibiting regulated secretion may or may not perform enzymatic modifications of secretory proteins in their secretory vesicles. If such modifications do occur, periodic induction with a secretagogue is necessary to retrieve active product, since the protein secreted constitutively is not fully processed and may thus be biologically inactive. If no modifications occur, the advantages of a controlled secretion process are limited to the (still important) relatively high product purity and concentration that can be achieved from the cell culture step.

For cycling operations to be justified when active protein is also secreted from unstimulated cells, the storage capacity of cells has to be large relative to basal secretion, so that a significant dynamic compression is achieved. In other words, the amount of protein secreted during the one to two hours of induced response should correspond to that synthesized over a period of at least 10 hours. Furthermore, basal secretion during recharging should be minimal, or most of the protein would be secreted in growth instead of secretion medium. Our evaluations with hGH-producing AtT-20 cells suggest that this cell line is not appropriate for production operations due to significant basal secretion during recharging.

If covalent modifications do occur in maturing secretory vesicles, the requirements of high storage capacity and minimal secretion during recharing are not strict because induction with a secretagogue is the sole way of retrieving active product. If the foregoing requirements are not met, quantitative suboptimalities apparently arise. We found that insulin-expressing AtT-20 cells exhibit good growth and secretion properties but are subject to limitations in the efficiency and extent of cycling operations. Some limitations (e.g. detachment of cells) may be overcome, but others (e.g. growth of foci of cells) may be intractable.

The βTC3 system offers a superior set of properties including high expression levels, stable cell growth, good secretion dynamics and efficient recharging. This is encouraging because it suggests that the problems encountered with AtT-20 cells are not inherent to transformed endocrine cell lines. This system shows potential for a practical production process, although it remains to be seen if foreign endocrine gene products can be expressed in these cells at high levels.

Besides endocrine hormones, cycling schemes might also be useful in the production of more complicated proteins expressed through recombinant DNA technology in cells exhibiting regulated secretion. It is unknown, however, whether proteins like tPA or Factor VIII will follow the regulated or constitutive pathway of secretion or whether their expression as fusion proteins with proper signal sequences will direct them into regulated sescretory vesicles. It may even by possible to use the proteolytic processing enzymes of the secretory granules to cleave the fused signal sequence at a dibasic linkage site, thereby yielding the natural product. Alternatively, those sequences might be used in separating the regulated proteins from the constitutively secreted ones by affinity chromatography: sorting proteins immobilized on columns bind specifically to the former class of proteins but not to the latter [25]. Further developments in the identification and characterization of sorting proteins and signal sequences should facilitate the proper expression of recombinant proteins in endocrine cells as well as the separation and purification of the polypeptide of interest. This, combined with the application of new methods (e.g. transgenic animals) for the development of cell lines with improved properties (e.g. βTC3) [19, 52], is expected to produce controlled secretion processes which are generic to a variety of complex products and which can harness the full potential of the regulated secretory pathway.

Acknowledgments: We would like to thank Dr. Harvey F. Lodish who contributed significantly to the direction of our research and whose numerous insightful suggestions have helped us bring this concept to fruition. The AtT-20 cells and βTC3 cells were graciously provided to H. F. Lodish by Regis Kelly and Shimon Efrat, respectively. This is a collaborative research project sponsored by the Biotechnology Process Engineering Center (BPEC) at M.I.T. under NSF research Grant No. CDR-8803014. G. E. Grampp was supported by an NSF fellowship.

6 References

1. Darnell J, Lodish H, Baltimore D (1986) Molecular cell biology. Scientific American Books, New York
2. Kelly RB (1985) Science 230: 25
3. Orci L, Ravazzola M, Amherdt M, Madsen O, Perrelet A, Vassalli J-D, Anderson, RGW (1986) J Cell Biol 103: 2273
4. Orci L, Ravazzola M, Storch M-J, Anderson RGW, Vassalli J-D, Perrelet A (1987) Cell 49: 865
5. Tooze J, Tooze SA (1986) J Cell Biol 103: 839
6. Mains RE, Cullen EI, May V, Eipper BA (1987) Ann NY Acad Sci 493: 278
7. Rassmussen H (1981) Calcium and cAMP as synarchic messengers Wiley, New York

8. Berridge MJ, Irvine RF (1984) Nature 312: 315
9. Schettni G, Florio T, Meucci O, Landolfi E, Cronin MJ, MacLeod RM (1987) Life Sciences 4: 2437
10. Wallis M, Howell SL, Taylor KW (1985) The biochemistry of the polypeptide hormones. John Wiley, New York
11. Wollheim CB, Biden TJ (1986) Ann NY Acad Sci 488: 317
12. Rajan AS, Hill RS, Boyd AE III (1989) Diabetes 38: 874
13. Seamon KB, Daly JWJ (1981) Cyclic Nucleotide Res 7: 201
14. Young LS, Naik SI, Clayton RNJ (1985) Endocrinology 107: 49
15. Moore H-PH, Kelly RB (1985) J Cell Biol 101: 1773
16. Moore H-PH, Walker MD, Lee F, Kelly RB (1983) Cell 35: 531
17. Sambanis A, Stephanopoulos G, Lodish HF (1990) Cytotechnology 4: 111
18. Sambanis A, Stephanopoulos G, Sinskey AJ, Lodish HF (1990) Biotechnol Bioeng 35: 771
19. Efrat S, Linde S, Kofod H, Spector D, Delannoy M, Grant S, Hanahan D, Baekkeshov S (1988) Proc Natl Acad Sci USA 85: 9037
20. Moore H-PH (1986) Ann NY Acad Sci 493: 50
21. Moore H-PH, Kelly RB (1986) Nature 321: 443
22. Copeland CS, Doms RW, Bolzau EM, Webster RG, Helenius A (1986) J Cell Biol 103: 1179
23. Dorner AJ, Bole DG, Kaufman RJ (1987) J Cell Biol 105: 2265
24. Munro S, Pelham HRB (1987) Cell 48: 899
25. Chung K-N, Walter P, Aponte GW, Moore H-PH (1988) Science 243: 192
26. Scammell JG, Burrage TB, Dannies PS (1986) Endocrinology 119: 1543
27. Hutton JC, Bailyes EM, Rhodes CJ, Rutherford NG, Arden SD, Guest PC (1990) Biochemical Society Transactions 18: 109
28. Winkler H, Fischer-Colbrie R (1987) Membrane receptors, dynamics, and energetics. Wirtz, KWA (ed) Plenum Press, New York
29. Eberwine JH, Jonassen JA, Evinger MJQ, Roberts, JL (1987) DNA 6: 483
30. Bienkowski RS (1983) Biochem J 214: 1
31. Halban PA, Wollheim CB (1980) J Biol Chem 225: 6003
32. Swabb EA, Wei J, Gullino PM (1974) Cancer Research 34: 2814
33. Lapp CA, Stachura ME, Tyler JM, Lee YS (1987) In Vitro 23: 686
34. Santerre R, Cook RA, Crisel RMD, Sharp JD, Schmidt RJ, Williams DC, Wilson CP (1981) Proc Natl Acad Sci USA 78: 4339
35. Sher E, Denisdonini S, Zanini A, Bisiani C, Clementi F (1989) J Cell Biol 108: 2291
36. Brunet-de Carvalho N et al (1989) Differentiation 40: 106
37. Tooze J, Tooze SA, Fuller SD (1987) J Cell Biol 105: 1215
38. Reid L (1988) presentation at the Engineering Foundation Conference on Cell Culture Engineering, Palm Coast, Florida, February, 1988
39. Sambanis A, Lodish HF, Stephanopoulos G (1991) Biotechnol Bioeng 38: 280
40. Bertrand G, Nenquin M, Henquin JC (1989) Biochemical J 259: 223
41. Grampp GE, Stephanopoulos G (1991) Controlled protein secretion in a pancreatic islet derived cell line. Presented at ACS National Meeting, Atlanta, Georgia, April 1991
42. Grampp GE, Lodish HF, Stephanopoulos G (1992) Paper in preparation
43. Wollheim CB, Pozzan T (1984) The Journal of Biological Chemistry 259: 2262
44. Adler M, Wong BS, Sabol SL, Busis N, Jackson MB, Weight FF (1983) Proc Natl Acad Sci USA 80: 2086
45. Sambanis A, Stephanopoulos G, Lodish HF, Sinskey AJ (1988) Engineering aspects of regulated protein secretion in animal cell cultures. Presented at ACS National Meeting, Los Angeles, California, September 1988
46. Dyken JJ, Vachtsevanos J, Sambanis A (1991) Protein secretion from endocrine animal cells: effects of chemical and physical stresses. Presented at ACS National Meeting, Atlanta, Georgia, April 1991
47. Morita S, Fernandezmejia C, Melmed S (1989) Endocrinology 124: 2052

48. Grampp GE, Sambanis A, Stephanopoulos G (1989) The characterization and use of regulated protein secretion in mammalian cell culture. Presented at ACS National Meeting, Miami Beach, Florida, September 1989
49. Franko AJ, Sutherland RM (1979) Radiation Research 79: 439
50. Park S, Singhvi R, Wang DIC, Stephanopoulos G (1990) Porous ceramic beads for animal cell culture. Presented at AIChE National Meeting, Chicago, Illinois, November, 1990
51. Applegate MA, Grampp GE, Stephanopoulos G (1991) Controlled protein secretion in a single-pass ceramic matrix bioreactor. Presented at ACS National Meeting, New York, New York, August 1991
52. Hanahan D (1989) Science 246: 1265

Biotechnological Reduction of CO$_2$ Emissions

I. Karube, T. Takeuchi and D. J. Barnes
Research Center for Advanced Science and Technology, University of Tokyo,
4-6-1 Komaba, Meguro-ku, Tokyo 153, Japan

Biotechnological fixation of carbon dioxide (CO$_2$) is described as a measure for reducing CO$_2$ emissions. Photosynthesis by microalgae would provide an efficient mechanism for the reduction of CO$_2$, if well-designed photobioreactors could be constructed for the intensive cultivations. Screening of microalgae which can grow well under high CO$_2$ concentrations would also be necessary in order to establish biotechnological CO$_2$ reduction systems. In addition, calcification and vegetation are discussed as mechanisms for reducing CO$_2$ emissions. Environmental monitoring is significantly important for the understanding of global CO$_2$ cycle, so that recent development in sensor technology are also described.

1 Introduction

Increasing atmospheric CO$_2$ has become a matter for serious concern because of the probable link between atmospheric CO$_2$ levels and global temperature — the greenhouse effect. There is no doubt that levels of CO$_2$ and other greenhouse gases have increased [1], and there is little doubt that the temperature of the northern hemisphere has risen by about 0.5 °C over the 100 years or so of instrumental measurements [2]. The 1980s were the warmest decade on record. It is not, as yet, certain that this increase in global temperature results from increased levels of greenhouse gases. However, it is not possible, nor does it seem wise, to dismiss the notion that global warming is already occurring.

The greenhouse effect describes a long-established, well-accepted theory of atmospheric science [3–4]. Greenhouse gases include CO$_2$, methane, chlorofluorocarbons and water vapor. These gases are relatively transparent to visible

Advances in Biochemical Engineering/
Biotechnology, Vol. 46
Managing Editor: A. Fiechter
© Springer-Verlag Berlin Heidelberg 1992

and near infrared radiation from the sun. Some of the received energy is re-emitted as infrared radiation. Greenhouse gases are gases which absorb in the infrared. The greenhouse theory supposes that the atmospheric concentration of greenhouse gases modulates global temperature by setting the balance between received and emitted energy.

CO_2 is the principal greenhouse gas. Atmospheric CO_2 has increased 25% in the past 150 years [5-7]. The increase in atmospheric CO_2 has been caused by burning of fuels, especially fossil fuels, in association with the increase in world population. Both of these have greatly accelerated over the time since the industrial revolution. The increase in population has not only increased the use of fuels, it has also led to considerable deforestation which has released CO_2 and reduced a significant sink for CO_2. If consumption of fossil fuels continues to show a 0.5-2,0% annual increase, atmospheric CO_2 concentration will reach twice the present level sometime next century [8-9].

It is difficult to predict with any certainty the increase in temperature that would be associated with a doubling of atmospheric CO_2 concentration. Present numerical models might be considered indicative rather than descriptive [10]. They agree very approximately at the global-scale but show wide differences at lesser scales. Many such differences occur because of other processes which are likely to affect global temperature and which are likely to be changed by global warming [11-13]. Modeling of such secondary processes is uncertain and feedback processes might serve to mitigate or to enhance a global temperature increase. Despite these uncertainties, present models suggest a global temperature increase of around 1.5-5.0 °C for a doubling in atmospheric CO_2 levels [14-17]. Such a temperature increase is likely to raise sea levels, to affect rainfall, to shift vegetation zones and to change world agriculture. These predictions have caused sufficient concern that many governments are considering ways to reduce CO_2 emissions. The Japanese Government, through the Ministry of International Trade and Industry (M.I.T.I.), is funding research into various mechanisms for reducing industrial CO_2 emissions. The initial goal for this research is development of procedures and processes which would significantly reduce CO_2 output from a thermal power station. Various chemical approaches to the reduction of exhaust CO_2 are being investigated. This article will focus on a possible alternative to chemical approaches; biotechnological reduction of CO_2 emissions. Figure 1 presents an overall concept for such a biotechnological system, as envisioned by M.I.T.I. [18].

2 Photosynthesis as a Technique for Reducing CO_2 Emissions

The principal cause of increasing atmospheric CO_2 levels is burning of fossil fuels with deforestation currently providing an annual increase in CO_2 which is about half that from fossil fuels. Both of these reservoirs of carbon were originally formed by photosynthesis. Consequently, photosynthesis provides the most obvious mechanism for biotechnological reduction of CO_2 in industrial exhaust gases.

Photosynthetic organisms produce organic compounds from CO_2 and H_2O using sunlight as an energy source. Photosynthetic fixation of CO_2 is accompanied

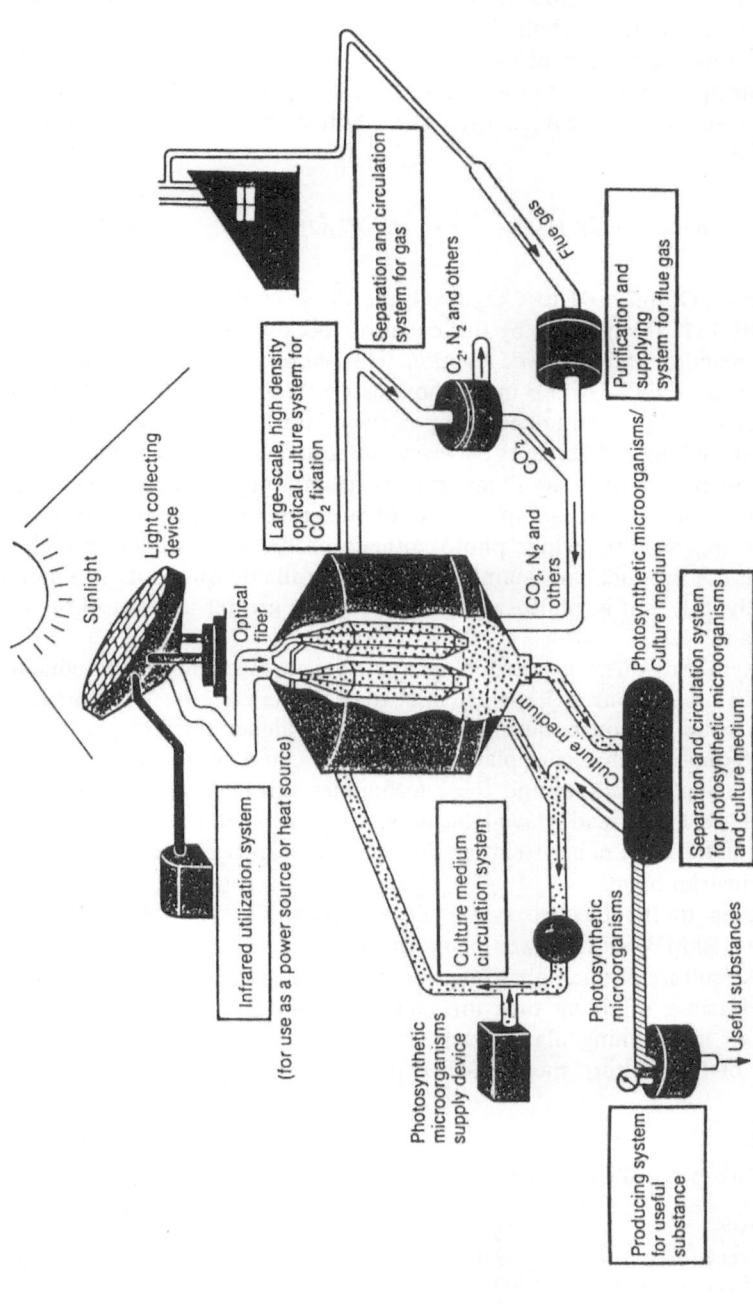

Fig. 1. A concept for biotechnological CO$_2$ fixation (Ref. 18)

by release of O_2. The process involves capture of light energy by photosynthetic pigments, the conversion of that energy to chemical energy stored as ATP and NADPH and the utilization of this chemical energy in metabolic syntheses; initially production of carbohydrates via the Calvin-Benson cycle. Release of O_2 occurs during conversion of light energy to chemical energy. Fixation of CO_2 occurs during subsequent synthesis of organic compounds. This synthesis depends upon the stored chemical energy. The photosynthetic process can be summarized as

$$n\,CO_2 + 2n\,H_2O \xrightarrow{\text{light}} n\,(CH_2O) + n\,O_2 + n\,H_2O .$$

The ratio of O_2 released to CO_2 fixed, the photosynthetic quotient [19], is not necessarily $1:1$, as suggested by this equation, because of the separation between the light reaction and the dark reaction. The photosynthetic quotient ultimately depends upon the $C:O$ ratio in compounds synthesized. The ratio of O_2 uptake to CO_2 evolved by a plant depends upon the relative rates of photosynthesis and respiration and upon differences between compounds synthesized and compounds respired. In plants, over the long-term, the respiratory quotient (CO_2 released: O_2 utilized) must reciprocate the overall photosynthetic quotient. Photosynthetic quotients approach 0.7 where photosynthetic products are mainly proteins and 1.3 where the products are mainly fats. Photosynthetic quotients in microalgae are usually >1, that is, photosynthesis produces a greater amount of O_2 than it consumes as CO_2.

Photosynthesis offers a possible, energy-efficient mechanism for reducing industrial CO_2 emissions. Microalgae appear to be the most likely agents for such photosynthetic fixation because they have much higher rates of photosynthesis per unit biomass than higher plants. Microalgae can be grown and manipulated using culturing techniques and these techniques can be adapted and scaled for industrial use. Using readily available data, it is possible to make crude, first order estimates of the scale of a system required to significantly reduce CO_2 production by an industrial plant.

According to the Electric Power Development Co. (6-15-1, Ginza, Tokyo, Japan), a 150 MW class power plant emits about $135\,t\,h^{-1}$ of CO_2 (Table 1). An intense culture of microalgae might contain 5 g dry weight per l. A very high, but still feasible, doubling time for such a culture might be 10 h. Continuous cultivation, maintaining algal density at $5\,g\,l^{-1}$, requires the introduction of $0.1\,l\,h^{-1}$ of freshculture medium. The outflow would then carry away excess

Table 1. Carbon Dioxide Emission for a 156 MW Thermal Power Plant

Power (MW)	156	Coal consumption $(t\,h^{-1})$	56.9
Heat efficiency (%)	38.0	Carbon Emission $(t\,h^{-1})$	37.0
Calories of coal $(kcal\,kg^{-1})$	6200	**CO_2 Emission $(t\,h^{-1})$**	**135.6**
Carbon content of coal (%)	65.0		

algae. If 50% of the dry weight is carbon then 1 l of such an algal culture would fix;

$$5 \, [\text{g l}^{-1} \text{ dry weight}] \times 0.1 \, [\text{l h}^{-1}] \times 50\% = 0.25 \text{ g h}^{-1} \text{ carbon}$$

$$= 0.25 \times (44/12) = 0.9 \text{ g h}^{-1} \text{ CO}_2 \,.$$

If this algal culture were used to fix the CO$_2$ released by a 150 MW power station (135 t h^{-1}) then the total volume of the culture would be

$$(135 \times 10^6 \times 0.9^{-1}) \times 1 \text{ l} = 15 \times 10^7 \text{ l} = 15 \times 10^4 \text{ m}^3 \,.$$

This estimate gives some idea of the area of ponds, or the size of a bioreactor, necessary to fix the CO$_2$ emitted by a small thermal power station. For example, if the algae were held in a 1 m deep pond, then the pond would need to be about 390×390 m. Here again, the estimate is indicative rather than definitive. The area value is probably conservative because optimistic algal production rates were used for the intial parameters. It is certainly conservative because it makes no allowance for self-shading likely in 1 m deep ponds containing such a high concentration of alga. Moreover, the value assumes constant illumination. A system entirely dependent upon natural light would require a very much larger pond to compensate for night-time respiratory release of some of the CO$_2$ fixed during the day. Furthermore, a system dependent upon natural light would require some mechanism for storage of the CO$_2$ emitted by the power station at night and during times of low natural light. It seems almost certain that photosynthetic reduction of CO$_2$ emissions will require some use of artificial light. This requirement suggests that a photosynthetic biotechnology will depend upon well-designed bioreactors rather than open pond systems.

3 Bioreactors for Intense Algal Cultivation

Rates of CO$_2$ fixation obtained with, essentially, single layers of photosynthetic organisms require vast areas of such organisms to fix the CO$_2$ released by a thermal power station. Microalgae offer economies of scale because they can be grown by volume instead of area. Further economies of scale can be obtained by growing algae in appropriately designed bioreactors instead of open ponds. It seems probable that such bioreactors would make maximum possible use of natural light because artificial light brings a penalty in additional CO$_2$ production.

The Okinawa Industry and Technology Promotion Association, Okinawa Prefecture, in southern Japan, developed a bioreactor for intensive cultivation of *Spirulina platensis* [20]. The system consisted of three 350 l reactor vessels which could be illuminated with natural and artificial light. Sunlight was collected by a heliostatic plane mirror and reflected onto a concave mirror (Fig. 2). The concave mirror focused the light onto the end of bundles of quartz optical fibers. These bundles transmitted the light to quartz rods immersed in the cultivation medium.

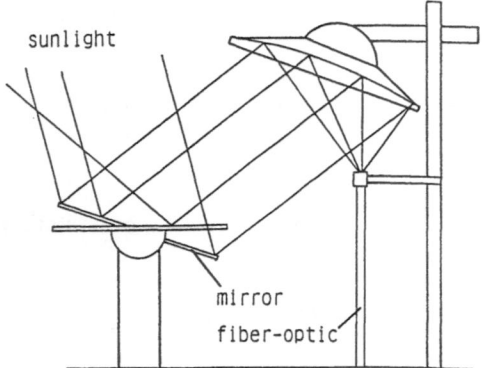

sunlight

mirror
fiber-optic

Fig. 2. Sunlight refraction collection system of a photobioreactor for growing *Spirulina platensis* (Ref. 20)

About 30% of the light collected was transmitted to the medium. Light from halogen lamps could also be transmitted to the medium using this system of mirrors and fiber optics. The cultivation medium also contained acrylic pipes housing fluorescent tubes which provided an additional source of light. The culture medium was aerated and agitated by bubbling with a mixture of air and CO_2. This bioreactor achieved a total production rate of $0.005 \text{ g l}^{-1} \text{ h}^{-1}$.

Samson and Leduy [21] reported multistage, continuous cultivation of *Spirulina maxima* in the laboratory using 4 flat tanks (1.22 m long × 0.51 m deep × 0.1 m wide) connected in series (Fig. 3). The tanks were illuminated by 16 Sylvania cool-white fluorescent tubes on each side. The cultures were aerated and agitated by bubbling them with air. The initial concentration of *S. maxima* was 0.4 g l^{-1}. The highest productivity obtained was $1.17 \text{ g l}^{-1} \text{ d}^{-1}$, or about $0.05 \text{ g l}^{-1} \text{ h}^{-1}$.

James and Al-Khars [22] reported 200 l bioreactors made of vertical tubes, 0.3 m in diameter (Fig. 4). The tubes were illuminated from the sides at a surface intensity of 330μ Einstein $\text{m}^{-2} \text{s}^{-1}$ and were bubbled with air containing 5% CO_2. A species of *Chlorella* was grown in this bioreactor. The system was monitored for a two month period during which maximum productivities approaching $0.3 \text{ g l}^{-1} \text{ d}^{-1}$ were achieved. This corresponds to a productivity $< 0.02 \text{ g l}^{-1} \text{ h}^{-1}$.

A much more successful experimental bioreactor for growth of *Chlorella pyrenoidosa* was demonstrated by Mori [23], although the volume of the reactor was small (Fig. 5). The bioreactor was equipped with a multitude of fiber-optic light radiators which gave a radiative surface area of 5800 cm^2 per liter of algal suspension. Visible light was focused onto the end of the fiber optic bundle using Fresnel lenses. Chromatic aberration in the lenses allowed visible light to be introduced into the algal suspension while reducing amounts of ultraviolet and infrared. The lenses were positioned so that green light was focused at the end of the fiber optic bundle. Most of the ultraviolet and 60% of the infrared were not, then, transmitted to the algal suspension. Xenon lamps were used to supplement natural light. The cultivation volume of the reactor vessel was 2.4 l. Optimum algal growth was obtained when algal suspensions were bubbled with a gas mixture containing CO_2 (8.5%), O_2 (10.4%) and N_2 (81.1%) at rates between 22 l h^{-1} and 214 l h^{-1}. Under these conditions, *C. pyreoidosa* grew about 8% per h in

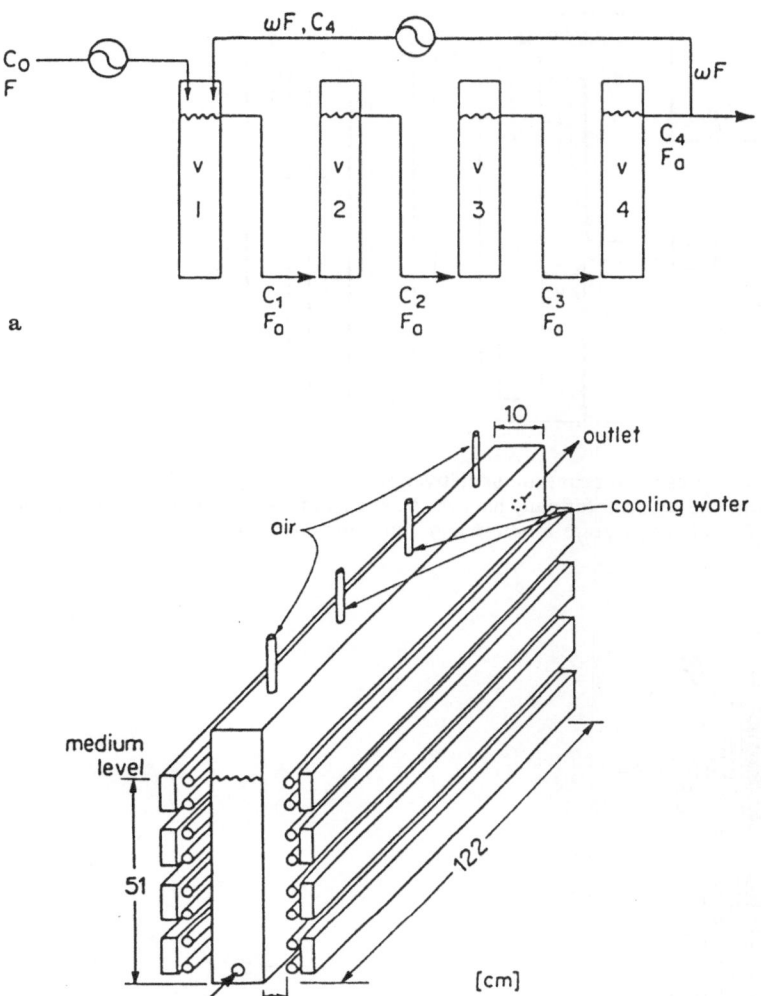

Fig. 3a, b. A flat tank photobioreactor for multistage, continuous cultivation of *Spirulina maxima* (Ref. 21). **a)** the complete system; **b)** detail of one photobioreactor

cultures containing 4–5 g dry weight per l. Algal growth was then around 0.3–0.4 g l^{-1} h^{-1} and the fixation rate approached 0.2 g l^{-1} h^{-1} CO$_2$. Doubling time of the algae was about 9 h. Algal growth rates obtained with this bioreactor were similar to those used above to estimate the likely size of systems which could fix the CO$_2$ produced by a thermal power station. Growth rates obtained with this bioreactor were up to 33 times higher than those reported for actual open pond systems containing *C. pyreoidosa* or similar microalgae.

Photobioreactors offer considerable advantages over open pond systems. It would be much easier to monitor and control the conditions in bioreactors than

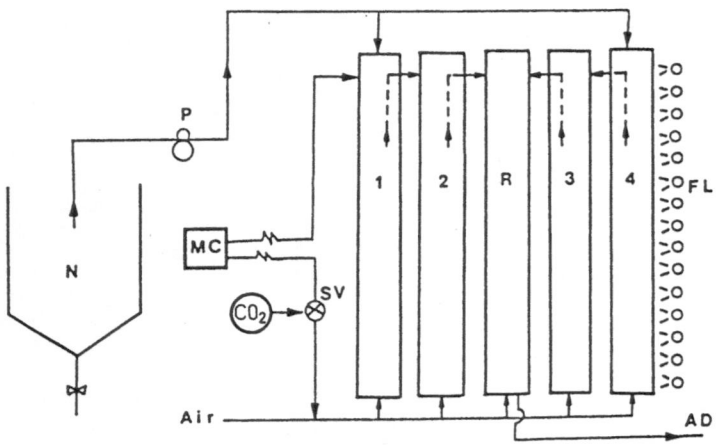

Fig. 4. A tubular bioreactor for continuous cultivation of *Chlorella* sp. (Ref. 22). *1 — 4*: tubular photobioreactors; *R*: reservoir for daily harvest; *N*: nutrient tank; *P*: pump; *FL*: fluorescent lights; *MC*: pH and temperature monitor and controller; *SV*: solenoid valve; *AD*: algal distribution pipe

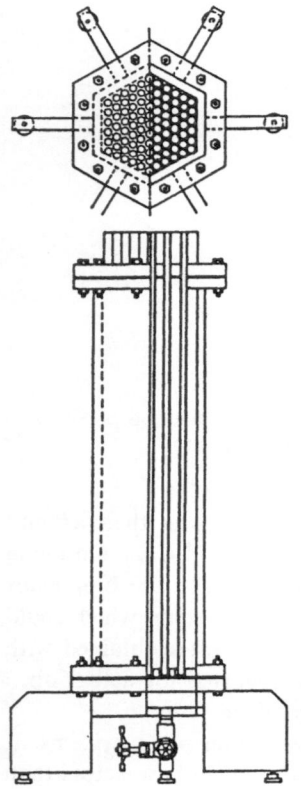

Fig. 5. A bioreactor tank for growing *Chlorella pyrenoidosa* (Ref. 23). Volume of algal suspension: 2.4 l; size of acrylic optical rod in a radiation tube: 3 mm diameter × 485 mm length; number of radiation tubes; 91; light radiative surface area: 5774 cm^2 l^{-1}

in open ponds. Thus, light, temperature, nutrients and CO_2 concentration could be manipulated within bioreactors to achieve optimal conditions for particular microalgae. Consequently, bioreactors offer possibilities for sustaining high rates of photosynthetic activity and algal growth. Moreover, bioreactors are far less susceptible to contamination and overgrowth by unwanted microorganisms than open ponds. It seems likely that efficient bioreactors for industrial use will combine artificial and natural illumination.

Bioreactors for controlling CO_2 emissions would produce very large amounts of biomass. Photosynthetic fixation of all of the carbon released from a 150 MW power station would generate nearly 2000 t of dry biomass per day. Consideration needs to be given to the fate of this biomass. Certain microalgae, including *Chlorella* and *Spirulina*, contain high levels of protein. Thus, industrial bioreactors based upon such algae might offset, or meet, their costs by sale of protein-rich products for use in agriculture or, perhaps, for direct human consumption. There will be considerable costs in running such bioreactors. The classic Redfield ratio suggests that production of 2000 t of dry biomass per day requires the daily addition of nutrients containing around 300 t of nitrogen and 20 t of phosphorus. Other elements likely to be present in the culture medium in significant amounts include Na, K, Ca, Mg, Fe, S, and Cl. It may be seen that biotechnology which has commercial potential would certainly aid and speed implementation of biological procedures for reducing industrial CO_2 emissions.

Biological reduction of CO_2 emissions will alleviate the problems associated with anthropogenic increase in atmospheric CO_2 only if the fixed CO_2 is somehow used to reduce burning of fossil fuels or, alternatively, if it is stored in a reservoir in which it has a long residence time (see below). Supplementation of boiler fuel for power stations is one such use. Given that a modern power station has an efficiency around 40% (Table 1), such procedures might significantly reduce the ratio of CO_2 emission for unit power generation.

Iwamoto [24] suggested that mass culture of *Botryococcus* sp. could provide a renewable source of hydrocarbons. *Botryococcus* is a green alga which uses hydrocarbons as storage products. Photosynthetic fixation of CO_2 into hydrocarbons may provide a mechanism for reducing CO_2 in power station exhaust gas. Hydrocarbons can constitute up to 60% of the dry weight of *Botryococcus*. Consequently, such procedures also offer another mechanism for significantly reducing CO_2 emitted for unit power generated. However, it would be more efficient, and it might be considerably more cost effective, to use such hydrocarbons to replace fossil hydrocarbons rather than to supplement power station boiler fuel.

Certain green algae, blue-green algae and phototrophic bacteria produce hydrogen. This area is currently of great interest [25, 26] because biologically-produced hydrogen could provide a clean, renewable fuel. Unfortunately, only very small amounts of hydrogen are generated by microorganisms. Hydrogen is generated in photosynthetic organisms when the energy trapped by the light reaction is not passed onto metabolic syntheses but is utilized in splitting water molecules. Thus, hydrogen is generated at the expense of carbon fixation and such organisms do not offer a simple, elegant means for reducing exhaust CO_2 emissions.

4 Microalgae which Tolerate High CO_2 Levels

Exhaust gases from thermal power stations typically contain 10–20% CO_2. This is more than 500 times the normal level of CO_2 in air. Thus, reduction of CO_2 emissions by microalgae will almost certainly depend upon species and strains which can tolerate unusually high levels of CO_2. Very little research has been carried out in this area.

The literature includes reports of mutant strains of microalgae which require, rather than tolerate, CO_2 levels up to 5% [27–29]. Such algae have provided a means for investigating mechanisms associated with CO_2 fixation. Marcus et al. [27] reported a high CO_2-requiring mutant of *Anancystis nidulans* R_2 that was isolated following treatment with the mutagenic agent *N*-methyl-*N'*-nitro-*N*-nitrosoguanidine. Similarly, Price and Badger [28] isolated high CO_2-requiring mutants of a blue-green alga, *Synechococcus* PCC7942, after treatment with ethylmethylsulphonate. It appears that, in both mutant strains, the algae were unable to convert HCO_3- to CO_2. Since the activity of ribulose 1,5-biphosphate carboxylase/oxygenase was not changed, it seems that the requirement for high CO_2 resulted from altered carbonic anhydrase activity. Ogawa (29) reported a mutant of *Synechocystis* PCC6803 which required high CO_2-levels for growth. In this case, the defect lay in the CO_2 transport system.

Algae and algal strains which can tolerate levels of CO_2 up to 20% have recently been isolated and identified in Japan in research efforts directly aimed towards

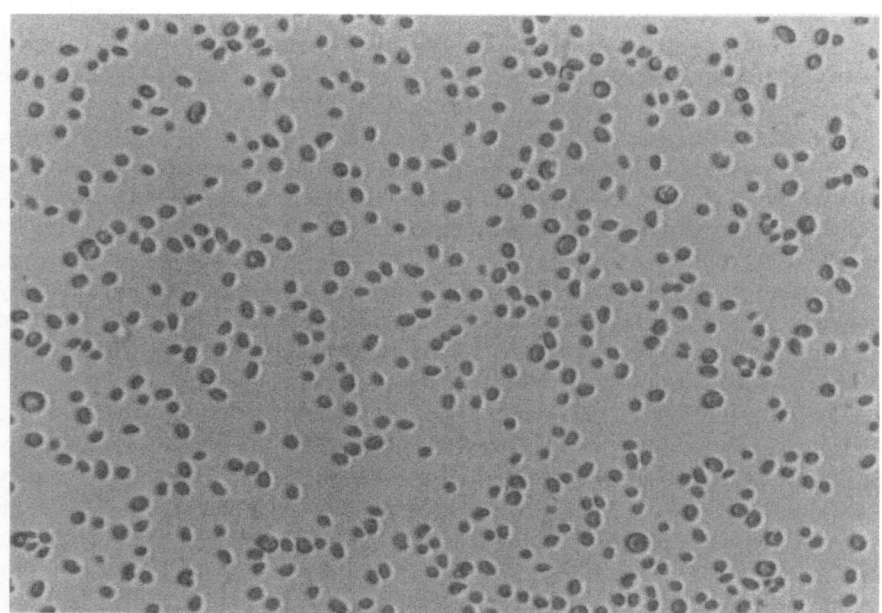

Fig. 6. *Oocystis* sp. Photomicrograph of a form tolerant to high CO_2 (× 400)

photosynthetic reduction of CO_2 levels in power station exhaust gases. Kodama and Miyachi [30] isolated a marine, green microalga, *Chlorococcum littorale*, which grew well at 30 °C and pH 4 in seawater bubbled with air containing 20% CO_2. Hanagata et al. [31] reported testing about 250 strains of algae for tolerance to high CO_2. They found several strains of green algae which could tolerate up to 20% CO_2. Takeuchi et al. [32] found that the green alga *Oocystis* sp. could survive and grow at CO_2 concentrations up to 20% (Fig. 6). The latter two groups are from our laboratory. Unpublished results obtained by these and other workers in our laboratory indicate that certain algal strains grow well in CO_2 concentrations up to 40% and at temperatures up to 40 °C. Given that work in this area began only recently, it seems probable that algae will be isolated which can tolerate extreme conditions while maintaining high growth rates.

5 Calcification as a Mechanism for Reducing CO_2 Emissions

CO_2 is principally removed from the atmosphere by two processes; photosynthetic transfer to the biosphere and weathering of rocks, especially carbonate rocks [33]. The biosphere (terrestrial and marine) contains about 4 times the amount of carbon in the atmosphere. Weathering of rocks ultimately leads to the addition of inorganic carbon to the sea. The sea provides a reservoir for CO_2 which is more than 50 times larger than the atmosphere. The principal flux of CO_2 between the atmosphere and the sea is due to gaseous exchange across the air-sea boundary. The sea does not represent a single, well-mixed reservoir. In essence, atmospheric CO_2 has direct access to only the surface layer of the sea — which may be considered as the layer above the thermocline. This reservoir is in equilibrium with the atmospheric reservoir and contains roughly the same amount of inorganic carbon [33].

Industrial production of CO_2 in 1985 was around 5×10^9 t [34]. This represents about 0.3% of the carbon stored in the terrestrial biosphere. Consequently, biological fixation of industrially produced CO_2 would require an annual increase of 0.3% in the terrestrial biosphere, or an annual increase of about 0.2% in the total biosphere. By any standards, this represents a huge additional biomass. It may be seen that biological reduction of CO_2 in industrial exhaust gases will have global significance only if the carbon is, eventually, transferred to a reservoir with a long residence time (the alternative is to use such biologically fixed cabon to replace fossil fuels — see above). Residence time is a function of the size of a reservoir and the rate of flux. The mean residence time of a molecule of CO_2 in the atmosphere is less than 10 years — much the same as the residence time for the biosphere. Residence time in the surface layer of the sea is perhaps 10–20 years and residence time in the deep sea is considered to be of the order of 1000 years [33]. Only the deep sea appears to offer a viable reservoir for long-term storage of excess CO_2.

Carbonate rocks store nearly 3×10^4 times more inorganic carbon than is in the atmosphere [33]. Given that vast carbonate deposits have accumulated since the Precambrian, the residence time for carbon in such rocks certainly exceeds

10 million years and perhaps approaches 100 million years. The bulk of carbonate rock is the result of biological precipitation of carbonates, mostly calcium and magnesium carbonates [35]. Clearly, calcification potentially offers a long-term biological solution for reduction of anthropogenic CO_2 emissions.

Possibilities for a biotechnology based upon calcification have not been explored. Calcification is a fundamental biological process having an importance to the living world which falls not far behind those of photosynthesis and respiration. It is then surprising that the metabolic processes leading to calcification (and ossification) are little understood. Because of this, there is very little information on which to base the development of such a biotechnology.

The fastest calcifying organisms are those associated with the formation of coral reefs — hermatypic scleractinian corals and calcareous red and green algae [36]. The hermatypic corals have a symbiotic relationship with a dinoflagellate alga. Calcification is rapid in corals and calcareous algae because photosynthesis somehow assists calcification in these organisms. Modern coral reefs are estimated to cover about 0.2% of the world's ocean area and about 15% of the shallow (<200 m) seas [37]. It has been estimated that coral reefs form a sink equivalent to about 2% of annual CO_2 emissions from burning of fossil fuels, and that this sink would be doubled by a moderate annual increase in sea level [38]. While coral reefs form a significant global sink for carbon, they are restricted to shallow tropical seas and there seems little possibility of their manipulation to provide a substantially larger sink. Moreover, coral reefs present a problem that exists with most other calcifying organisms — the precipitation of calcium carbonate generates protons that must be removed or neutralized for precipitation to continue. At its very simplest, this could be represented as

$$Ca^{2+} + CO_2 + H_2O \longleftrightarrow CaCO_3 + 2H^+.$$

Studies indicate that calcification in almost all organisms lowers the pH of the surrounding medium. The process of calcium carbonate formation in the ocean changes the seawater carbonate equilibrium such that $p CO_2$ is raised [39]. This can cause loss of CO_2 to the atmosphere. For instance, each mole of carbonate precipitated from seawater above the Bahama Banks was accompanied by the loss from the system of about 0.6 mol CO_2 [40]. About half this loss could be accounted for as evasion through the sea surface to the atmosphere. Although calcification may cause short-term, localized release of CO_2 from seawater to the atmosphere, the overall effect is to transfer CO_2 to a large reservoir which has a long residence time.

The simplest and oldest theory of algal calcification is that photosynthetic removal of CO_2 displaces the carbonate equilibria in freshwater and seawater [41];

$$H_2O + CO_2(aq) \longleftrightarrow H_2CO_3$$
$$\longleftrightarrow H^+ + HCO_3^- \longleftrightarrow 2H^+ + CO_3^{2+},$$

increasing the carbonate concentration. This then causes the solubility product for calcium cabonate to be exceeded, leading to precipitation. This precipitation reaction depends upon high pH and carbonate concentration being maintained. It seems possible that CO_2 from exhaust gases could be used to dissolve calcium carbonate in much the same way that atmospheric CO_2 dissolved in rain assists weathering of carbonate rocks. If algae could be found which can tolerate high pH and carbonate, then photosynthesis by such algae may lead to precipitation of calcium carbonate. If the precipitation reaction is stoichiometric for carbon, in the manner described by Goreau (42) for light-enhanced calcification in corals;

$$Ca^{2+} + 2\,HCO_3^- \quad \longleftrightarrow \quad CaCO_3 + H_2O + CO_2\,,$$

then the process would not store exhaust CO_2 as carbonate. It would present a potentially useful method for transporting and storing CO_2 for subsequent photosynthetic fixation. If, on the other hand, excess Ca^{2+} could be supplied, say from seawater, and if conditions can be established such that precipitation of carbonate does not require photosynthetic removal of an equivalent amount of CO_2, then exhaust CO_2 could be stored as carbonate. Such a carbonate solution-precipitation system offers a number of advantages in the development of industrial processes for reduction of CO_2 in industrial exhaust gases, regardless of its ability to store exhaust CO_2 as carbonate.

6 Increasing CO₂ Fixation by Vegetation

There is more than twice as much carbon stored in terrestrial vegetation as is present in the atmosphere. Moreover, there is an annual oscillation in atmospheric CO_2 concentrations which is caused by summer growth of vegetation in the more extensive, more vegetated land masses of the northern hemisphere. Thus, terrestrial vegetation provides a considerable reservoir for carbon. Anthropogenic changes to vegetation are significantly diminishing this reservoir and releasing about 2.5×10^9 t of CO_2 to the atmosphere each year. This represents about half the annual release of CO_2 from fossil fuels to the atmosphere. While increasing worldwide urbanization contributes to this vegetative change, the overwhelming effect is due to continued decrease in the amount of tropical rainforest. About 1% of the total area of tropical rainforest, around 10^5 km², is lost each year. Much of this loss is due to highly inefficient "slash and burn" agricultural practices. Because of the nature of tropical soils, the cleared land is usually productive for only a few years and must then be replaced by destruction of more rainforest. Cleared but now useless land does not quickly return to tropical rainforest since this climax of vegetative growth involves ecological processes taking hundreds of years. Tropical rainforest is mainly found in developing nations. Assistance from developed nations in agriculture and land practice might halve the rate of loss of tropical rainforest [43].

Conservation of existing tropical rainforest is an urgent requirement. Possibilities exist for significant expansion of other vegetation. Unfortunately, such expansion would not provide a significant, long-term reservoir for anthropogenic carbon. The average annual fixation of CO_2 by vegetation is estimated at 650 t per km². Fast growing trees may form mature forest in 25–35 years [43]. Consequently, such forest would lock away around 2×10^4 t of CO_2 per km^2 before reaching an equilibrium between fixation and loss. It is estimated that afforestation of 10% of the world's savanna and 20% of pampas and prairie would fix around 1.5×10^9 t of CO_2 annually [43]. Success with this enormous agricultural task would only remove around 20% of the annual anthropogenic CO_2 production – and such removal would continue for about 3 decades before this reservoir moved towards equilibrium. Afforestation of the whole of the Sahara Desert would fix CO_2 equivalent to the annual output from fossil fuels [43] but, unless the timber is somehow used to replace fossil fuel use, an equivalent area would need to be afforested within a few decades. Greening of deserts provides a highly desirable goal – but more because it would create productive land to support an increasing population than because it would provide a significant, long-term sink for anthropogenic CO_2.

7 Monitoring of Environmental CO_2 Levels

Better understanding of the global CO_2 cycle requires monitoring of environmental levels of CO_2. Such scientific investigations are possible with present technology, including satellite technology. Present instrumentation for monitoring environmental CO_2 is complicated and difficult to use. There is a great need for a small, simple, robust instrument which could be used for routine monitoring of CO_2 levels by industry and, perhaps, by environmental inspectors.

Present sensors for CO_2 have poor sensitivity and long response times. They surround the glass membrane of a pH electrode with a solution of sodium bicarbonate [44–45]. The bicarbonate solution is held within a gas-permeable membrane, usually a silicon membrane. Diffusion of CO_2 through the membrane into the bicarbonate solution is proportional to the partial pressure of CO_2 in the surrounding environment. Changes in the amount of CO_2 alter the pH of the bicarbonate solution and the pH electrode responds to these changes. It is because this type of sensor depends upon a Nernstian-type response that the sensitivity is low and response times are long. Improved response times have been obtained by miniaturization and improved electrode geometry.

Amperometric sensors usually give a higher sensitivity than potentiometric sensor because a current generated by electrochemical reactions is directly proportional to the concentration of target compounds. Karube et al. [46] devised an amperometric biosensor for CO_2 based upon a Clark-type oxygen electrode and an immobilized, autotrophic bacterium which consumes CO_2 and releases O_2 (Fig. 7). At pH 5.5 and 30 °C, a linear response curve was obtained in the range between 5 and 200 mg l^{-1}. The sensor response was stable for a month. The thermostability of the sensor was improved by the use of a thermophilic bacterium that was collected from Atagawa hot spring (approximately 80°C) in Shizuoka, Japan [47]. The sensor's operation range was 34 to 58 °C, and a linear relationship was obtained in the range 1 mM to 8 mM $NaHCO_3$ and 3% to 12% CO_2 in air. These sensors have been miniaturized using semiconductor fabrication techniques

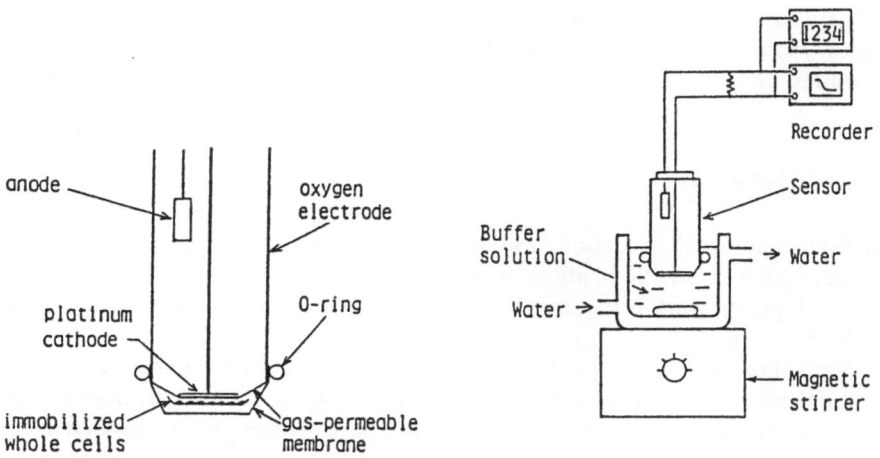

Fig. 7. CO_2 sensor based upon an autotrophic bacterium (Ref. 46). The bacterium converts CO_2 to O_2 which is then monitored with a Clark-type O_2 electrode

[48]. The response time was 2 to 3 min and the response was linear in the range 0.5 to 3.5 mM $NaHCO_3$. Figure 8 shows the typical structure of such sensors.

Other forms of CO_2 sensor have been explored. For example, Yamazoe et al. developed CO_2 sensors based upon a combination of a solid electrolyte (NASICON) and a Li_2CO_3-$CaCO_3$ electrode [49]. The sensors gave a good Nernstian-type response under wet conditions; the 90% response time for 250 ppm CO_2 was approximately 8 seconds.

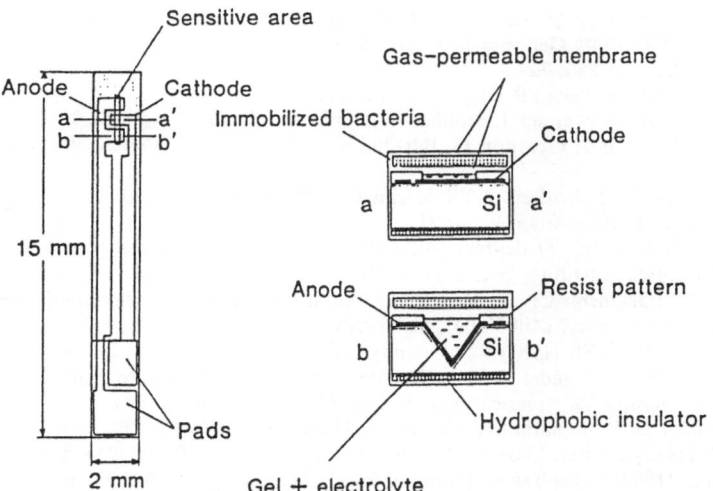

Fig. 8. Bacterial CO_2 sensor constructed using semiconductor fabrication techniques (Ref. 48)

There exsits a considerable medical and safety market for an improved CO_2 sensor. However, the necessity for a CO_2 sensor for environmental monitoring is becoming ever more urgent.

8 Conclusions

(a) Biotechnological reduction of industrial CO_2 emissions will probably depend upon photosynthetic fixation by microalgae. Considerable experimental and development work is required before bioreactors suitable for such fixation can be designed.

(b) Such photobiotechnological processes will consume large amounts of inorganic nutrients and produce very large amounts of organic materials. Sources for the inorganics and the fate of the organics have scarcely been considered.

(c) Photobiotechnological reduction of CO_2 emissions will only be effective in countering the rise in atmospheric CO_2 if the fixed carbon replaces fossil fuels or is transferred to a long-term reservoir. The deep ocean and carbonate rock offer two such reservoirs.

(d) Conservation of existing terrestrial vegetation would significantly reduce the rate at which atmospheric CO_2 levels are presently rising. However, vegetation of presently barren land and revegetation of denuded land would not offer long-term solutions to the problem of rising atmospheric CO_2.

(e) There is a fundamental requirement for better understanding of the CO_2 cycle which demands better instrumentation for global monitoring of CO_2.

9 References

1. Neftel A, Moor E, Oeschger H, Stauffer B (1985) Nature 315: 45
2. Hansen J, Lebedeff S (1988) Geophys Res Lett 15: 323
3. Schneider (1987) Sci Am 256: 72
4. Kasting JF, Toon OB, Pollack JB (1988) Sci Am 257: 90
5. Neftel A, Oeschger H, Schwander J, Stouffer B, Zumbrunn R (1982) Nature 295: 220
6. Jouzel J, Lorius C, Petit J, Genthon C, Barkhohh N, Katolyoff V, Petrov V (1987) Nature 329: 403
7. Barnola JM, Raynus D, Korotkevich YS, Lorius C (1987) Nature 329: 408
8. Edmonds JA, Reilly J (1984) Energy J 4: 21
9. Mintzer IN (1987) A matter of degrees: the potential for controlling the greenhouse effect. World Resources Institute, Washington, DC
10. Washington WM, Parkinson CL (1986) An introduction to three-dimensional climate modeling. University Science, Mill Valley, California
11. Schlesinger M, Mitchell JFB (1987) Rev Geophys 25: 760
12. Hansen JE, Takahashi T (eds) (1984) Climate processes and climate sensitivity: Geophysical Monograph 29. American geophysical Union, Washington, DC
13. Cess RD, Hartman D, Ramanathan V, Berroir A, Hunt GE (1986) 24: 439
14. Clark WC (ed) (1982) Carbon Dioxide Review 1982. Oxford Univ Press, New York
15. Pearman GI (ed) (1987) Greenhouse: Planning for climate change. Brill, Leiden
16. National Research Council (1987) Current Issues in Atmospheric Change. National Academy Press, Washington, DC

17. Bolin B, Doos BR, Jaeger J, Warrick RA (eds) (1986) The greenhouse effect, climatic change and ecosystems. Wiley, New York
18. Kuwahara H (1991) Sci & Tech Japan 10: 14
19. Rhyther JH (1956) Limnol Oceanogr 1: 72
20. Machinery System Development Association (1988) Report of feasibility studies on the development of blue-green algae production system (in Japanese)
21. Samson R, Leduy A (1985) Can J Chem Eng 63: 105
22. James CM, Al-Khars AM (1990) Aquaculture 87: 381
23. Mori K (1986) Biotechnol Bioeng Symp 15: 331
24. Iwamoto H (1990) Cultivation of *Botryococcus braunii* and hydrocarbon production. In: Miyachi S, Karube I, Ishida Y (eds) Current topics in marine biotechnology, Fuji Technology Press, Tokyo, p 123
25. Kumazawa S, Mitsui A (1989) Algae for hydrogen generation. In: Kitani O, Hall CW (eds) Biomass handbook, Gordon and Breach Scientific, New York, p 219
26. Miyake J, Asada Y, Kawamura S (1989) Nitrogenase. In: Kitani O, Hall CW (eds) Biomass handbook, Gordon and Breach Scientific, New York, p 362
27. Marcus Y, Schwarz R, Friedberg D, Kaplan A (1986) Plant Physiol 82: 610
28. Price GD, Badger MR (1989) Plant Physiol 91: 514
29. Ogawa T (1990) Plant Physiol 94: 760
30. Kodama M, Miyachi S (1991) Abstract, p 73: International Marine Biotechnology Conference '91. Baltimore
31. Hanagata N, Takeuchi T, Fukuju Y, Barnes DJ, Karube I (1992) Phytochemistry, in press
32. Takeuchi T, Utsunomiya K, Owada M, Karube I (1992) J Biotechnol, in press
33. Skirrow G (1975) The dissolved gases − carbon dioxide. In: Reley JP, Skirrow G (eds) Chemical oceanography, vol 2, Academic, London, p 1
34. Special research group for global environment, National Research Institute for Pollution and resources (ed) (1990) Global warming: Its technological countermeasure. Ohm-sha, Tokyo
35. Milliman JD (1974) Marine carbonates, Springer, Berlin Heidelberg New York, 375 pp
36. Barnes DJ, Chalker BE (1991) Calcification and photosynthesis in reef-building corals and algae. In: Dubinsky Z (ed) Coral reef ecosystems, ecosystems of the world, No 24, Elsevier, p 109
37. Smith SV (1978) Nature 273: 225
38. Kinsey DW, Hopley D (1991) Paleogeogr Paleoclimat Paleoecol 89: 363
39. Garrels RM, Mackenzie FT (eds) (1981) Carbon dioxide research and assessment program: some aspects of the shallow ocean in global carbon dioxide uptake. US Dept energy, CONF-8003115 UC11, 83 pp
40. Broecker WS, Takahashi T (1966) J Geophys Res 71: 1575
41. Borowitzka MA (1982) Mechanisms in algal calcification. In: Round FE, Chapman DJ (eds) Progress in Phycological Research, I, Elsevier Biomedical, Amsterdam, p 137
42. Goreau TF (1959) Biol Bull 116: 59
43. Karube I (1991) Sci & Tech Japan 10: 9
44. Stow RW, Baer RF, Randall BF (1957) Arch Phys Med Rehabil 38: 646
45. Severinghaus JW, Bradley AF (1958) J Appl Physiol 13: 515
46. Suzuki H, Tamiya E, Karube I (1987) Anal Chim Acta 199: 85
47. Suzuki H, Tamiya E, Karube I, Oshima T (1988) Anal Lett 21: 1323
48. Suzuki H, Kojima N, Sugama A, Takei F, Ikegami K, Tamiya E, Karube I (1989) Electroanalysis 1: 305
49. Miura N, Yao S, Shimizu Y, Yamazoe N (1991) Digest of Technical Papers, Transducers '91, San Francisco, p 558

Immunochemically Based Assays for Process Control

B. Mattiasson and H. Håkanson
Department of Biotechnology, Chemical Center, Lund University,
P.O. Box 124, S-22100 Lund, Sweden

Immunoanalysis for process control is reviewed. The development of non-equilibrium immunoassays from a manually operated flow injection system to a completely computerized system for sample analysis as well as calibration and data evaluation is discussed. Several other approaches in using immunochemistry in biosensors for process control are discussed. The fact that reuse of antibodies raises specific demands on their properties is discussed. Even if the area is young, much progress has been achieved and one can foresee an interesting future development of the area.

Abbreviations

FIA: Flow Injection Analysis
ELISA: Enzyme Linked ImmunoSorbent Assay
IgG: Immunoglobulin G
ConA: Concanavalin A

Advances in Biochemical Engineering/
Biotechnology, Vol. 46
Managing Editor: A. Fiechter
© Springer-Verlag Berlin Heidelberg 1992

1 Introduction

More detailed knowledge about a process should make it possible to control that process better. In the field of biotechnological processes, the development of good and efficient ways of collecting appropriate data has been slow. The same is true for the development of control strategies to utilize the collected data for process control. Much information is still lacking about the modelling of many processes, and the development of sensors for specific monitoring is only in its infancy.

After the first description of a biosensor by Clark and Lyons [1], much work has been done in this area and numerous publications have appeared. However, only a few have been developed to the point where they can withstand the harsh conditions which apply in practical applications. The main bottlenecks in development are the robustness of the sensor, sampling and sample handling, and the availability of appropriate knowledge of metabolism and control strategies to deal with metabolism.

A biosensor is presently defined as a man-made sensing device constructed by combining the specificity of biomolecules with the signal transducing and processing capability of competent electrical and/or optical components. In this area most interest has focused on enzyme-based sensors and their use in the quantification of low molecular weight molecules [2]. Most work using biosensors for process control has been carried out in research laboratories [3, 4], but there are examples of their use in bioprocess monitoring [5–7]. Even less has been done in the field of immuno-based sensors, in spite of the fact that there have been reports about the potential of such an approach [8]. The driving force behind development of such sensors is the desire to monitor the concentration of macromolecules in bioprocesses frequently or continuously. This would make it possible to register and document the processes; it would also be valuable for validation of processes. Analytical data could be used to control the process. Initially, deeper understanding and better documentation seem to be the primary driving forces. The possibility of registering concentrations of interesting compounds in a reaction will help to deepen the basic understanding of the process.

Immunoanalysis is traditionally regarded as having some inherent properties making it less suitable for process monitoring. In Table 1, some characteristics of traditional enzyme-based and immuno-based analyses are listed. It is clear that there are many reasons to develop immunosensors, since it would make it possible to specifically monitor macromolecules and to detect small molecules for which no suitable enzyme reactions are available. However, the time required for an assay has been a strongly limiting factor in the application of immunoanalysis. In most cases immunochemical assays are performed over several hours or overnight. With the introduction of homogeneous immunoassays it was clearly shown that it is possible to operate an immunoassay with an instantaneous reading [9–14]. In the EMIT-concept (Enzyme Multiplied ImmunoTechnique), the immunological binding reaction is read spectrophotometrically within seconds of mixing the reagents [11]. It has been adopted for use in centrifugal analyzers with extremely short time periods between mixing and reading. Thus, there is no inherent

Table 1. Comparison of enzymatic and immunochemical analyses

	Enzyme/Substrate	Antibody/Antigen
Operational range mol^{-1}	10^{-6}	10^{-12}
Specificity	+	+ +
Availability for new analyses	−	+
Ease of producing new reagents	−	+
Usefulness for quantitating macromolecular structures	− −	+
Time needed	+	−

property of the immunochemicals that requires that the assays should be operated over several hours. One characteristic that is obvious in analytical data from immunochemical binding assays is the variation in the data: $\pm 10-15\%$ is not unusual. This is ascribed to the many manual steps in such an assay and the fact that in order to shorten the procedure, each step is not allowed to reach equilibrium. To interrupt any step earlier than equilibrium has been reached is sensitive and requires a very reproducible mode of operation.

Thus, when designing an immunoanalysis to be used for process control such an assay must have the following characteristics:
− quick
− sensitive
− very reproducible in operation
− stable.

With this in mind we started to develop quick and reliable immunoassays about fifteen years ago, with the ultimate goal of designing immunosensors for process control.

This chapter summarizes some of this development and covers some recent aspects of the use of immunosensors for monitoring and control in down stream processing.

There are two different principles used in the construction of an immunosensor: either the direct binding interaction between antigen and antibody is monitored, or the binding reaction is measured in the presence of some additional chemicals. The latter is used in the immunoassays that are presently used in clinical chemistry and it may therefore be quicker to develop a sensor based on these principles. The former, the direct interaction, holds great and interesting possibilities, but more basic work is needed to develop this aspect. Both these aspects will be discussed in some depth in this paper.

2 Binding Assays Involving Consumption of Reagents

Competitive binding assays, as well as sandwich assays, require the addition of reagents prior to the reading of the result of the assay. This means that it is difficult and practically impossible to operate such assays in situ in bioreactors. Thus, such

Table 2. Examples of analyses performed and detectors used in flow injection binding assays

Target molecule	Detector	Ref.
Human serum albumin	Calorimeter	[15]
Transferrin	Spectrophotometer	[35]
IgG	Electrochemical detector	[36]
IgG	Electrochemical detector	[37]
Human IgG	Nephelometer	[38]
Human IgG	Ellipsometer	[32]
Human IgG	Spectrofluorimeter	[39]
Gentamycin	Spectrophotometer	[16]
Insulin	Polarographic oxygen electrode	[40]
Proinsulin	Calorimeter	[41]
Methyl-α-mannopyranoside	Spectrophotometer	[17]
Theophyllin, insulin	Ion selective electrode	[42]
17-α-Hydroxyprogesterone	Photomultiplier	[43]

assays are operated with sampling of a side stream of the reaction mixture while performing the assay off-reactor. The assay must still conform to the characteristics listed above if it is to be useful for process control.

The combination of immunochemical reactions with flow injection analysis has become increasingly popular. Several applications have been published based on this combination (Table 2). Flow injection analysis is usually characterized by high reproducibility of contact time, sample volumes, etc., features of great importance when designing immunochemical binding assays.

2.1 Non-Equilibrium Immunoassays

In the development of competitive immunological binding assays the outcome of the analysis is highly dependent upon the reproducibility of the incubation steps. In order to improve this, antibodies were immobilized to solid particles and packed in a column [15, 16]. This packed bed was placed in a continuous flow of buffer and reagents were introduced as a pulse into the flow. Such manually operated immunoassays showed that it was possible to operate far from equilibrium and maintain high reproducibility. In fact, reproducibility was better than that reported for conventional assays.

A typical assay was set up as schematically shown in Fig.1.

A critical point in this context is the dissociation step in which the affinity sorbent is rinsed and restored for the next cycle. It is important to achieve the most complete dissociation possible, since any remaining reagents will reduce the capacity of the column. Remaining labelled antigen may also interfere with the subsequent assay. The dissociation step is time dependent, and a compromise must be reached between the amount of time consumed in dissociation and the amount of undissociated material remaining on the support. In Fig. 2 the

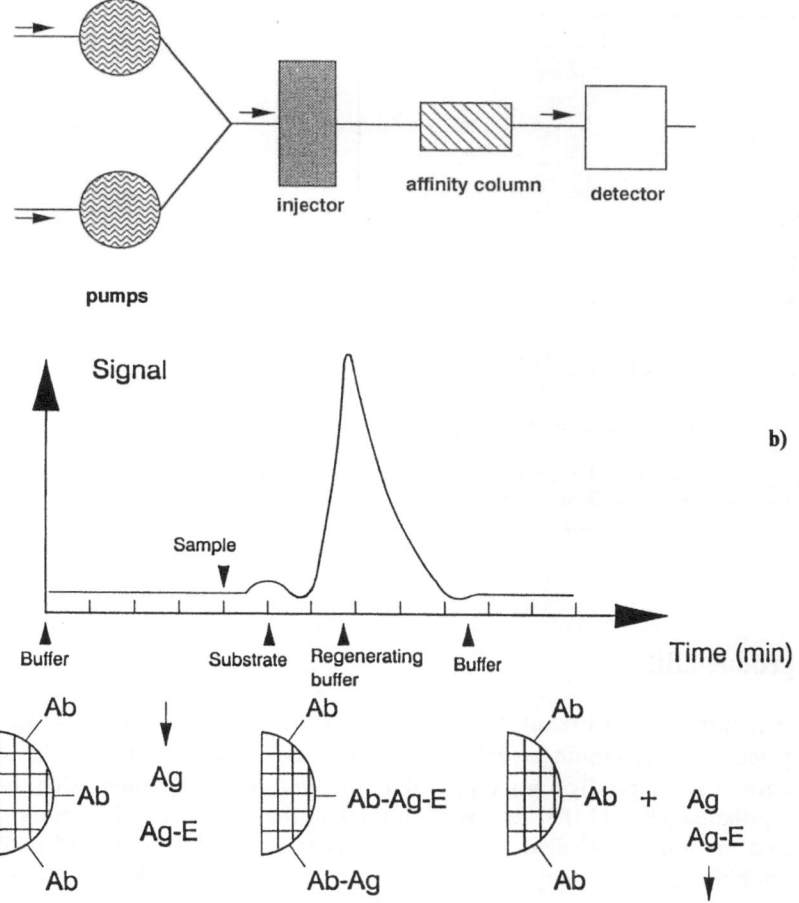

Fig. 1a. Schematic presentation of the experimental set-up for a manual flow injection-ELISA. **b)** Schematic presentation of an assay cycle in the flow-ELISA. At the bottom is shown the immunosorbent during the cycle. The time scale indicates time periods needed for the different steps in a typical assay cycle. Additions of the different liquids is done at the *arrows* indicated

relationship between the remaining enzyme activity and the regeneration time for a column with immobilized anti-transferrin antibodies with bound transferrin ånd horseradish-peroxidase-labelled transferrin is shown. The exact levels of substance remaining on the column after washing are dependent on the molecular characteristics of the system used in each case.

Using this kind of experimental set-up, several transducers were evaluated. Initially most of the work was done using calorimetric detection, but it soon became evident that it is more convenient to operate with either electrochemical or spectrophotometric detection.

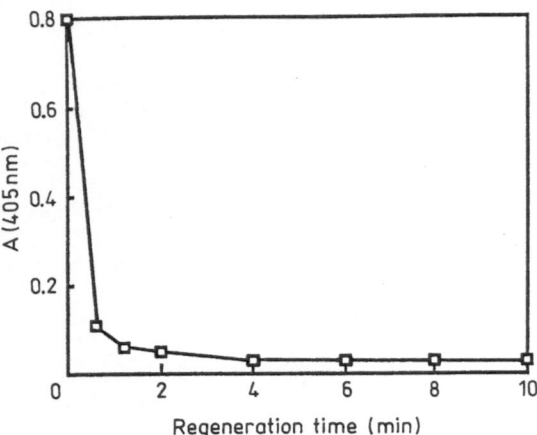

Fig. 2. Remaining activity of the labelling enzyme left on the immunosorbent after an assay cycle including washing for different time periods. (The data were obtained from a transferrin assay using horseradish peroxidase as labelling enzyme)

2.2 Reproducibility

When the immobilized adsorbent is reused it is essential to maintain good control of the efficiency of the binding reaction. The capacity of the adsorbent column may be reduced substantially upon repeated use without markedly influencing the analytical outcome (Fig. 3). This is based on the fact that the sample is premixed with a fixed amount of labelled antigen and this mixture is then subjected to competitive binding to the immunosorbent. Even if a reduction in the capacity of the immunosorbent occurs, the competitive situation is unchanged. Provided the sites on the support are regarded as equal, then the competitive binding is constant.

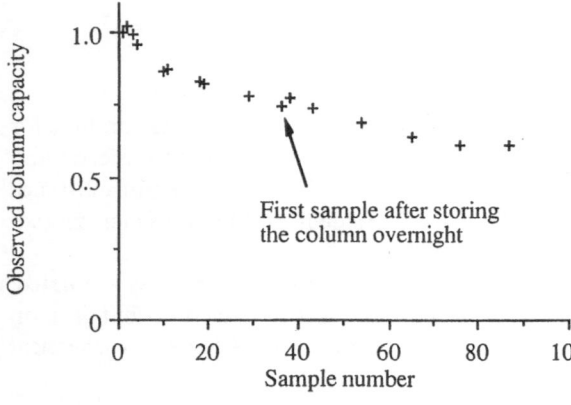

Fig. 3. Reduction in capacity of the immunosorbent as a function of the number of assay cycles

It should be stressed in this context that the reuse of the immobilized antibodies raises some new constraints. It is difficult to efficiently dissociate the bound antigen if high affinity antibodies which bind the antigen strongly are used. It may be possible to achieve efficient dissociation by a lengthy procedure, but this is not acceptable when designing a rapid process. Two options exist: either the use of harsher conditions, or to use weaker antibodies. If the former is used, it may be possible to obtain efficient dissociation for a few cycles, but the rate of denaturation will increase dramatically, and the life-span of the adsorbent will be reduced. A weaker antibody will lower the sensitivity of the assay, but will also make it possible to dissociate the bound antigen in a quick and efficient manner. Experiments showed that the loss in sensitivity was not dramatic. It was possible to measure concentrations down to 10^{-9} moles l^{-1} within the time frame given in Table 2. If higher sensitivity was needed, a slower assay cycle was used: then, the sensitivity was pushed almost four orders of magnitude downwards. In most cases, however, a biotechnological process yields higher product concentrations, and thus the faster version of the assay is sufficient. In addition to analyzing the concentration of the product of interest it may be useful to measure the concentration of specific impurities present in the sample.

2.3 Developments Towards the Use of Flow-Injection Immunoassays for Process Control

2.3.1 Computerized Flow-Injection Immunoassays

After establishing the performance characteristics of the manually operated systems the assay was automated. From this step to the use of the system for process control is a small but significant step.

A computerized system was first used in data collection and evaluation [17], and was later also used for controlling pumps and valves [18, 19].

Data acquisition and evaluation can be done in different ways. The most primitive way was utilized in the manually operated assays. The signals were registered on a recorder and the peak amplitudes were measured using a ruler. This procedure was accurate, but was difficult in cases of fluctuating base-line. To improve the speed of the assay, the peak was registered in the computer as a data signal every second, and the data were stored for later evaluation. It is not necessary to read the whole peak, as long as the same region is studied in each assay. A point on the tail of a peak from a traditional FIA-registration represents a more dilute sample than the peak value. This is reproducible from incubation to incubation.

In the computerized evaluation of the assay peak from the enzyme-catalyzed reaction of the labelling enzyme captured on the immunosorbent, proper and reproducible analysis of a small fraction of the tailing peak gives all the information needed. This is one of the key features of flow injection analysis — the peak contains much more information than just the peak height [20]. Thus, it is not necessary to run the whole analytical step as a separate unit operation

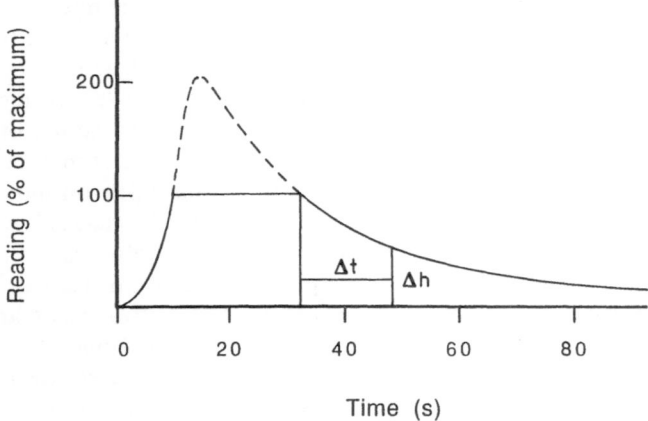

Fig. 4. A typical peak from an experiment with substrate included in the buffer flow. The enzymatic reaction normally produced peaks out of range of the photometer which is represented by the *dashed portion* of the curve. For a manual evaluation of the curves, a measurement (Δh) was made at a fixed time (Δt) after the curve passed the 100% limit. (From Ref. 17, with permission)

in the assay procedure: the same sampling time can be reduced further in order to save incubation time. The most extreme case is illustrated in Fig. 4 where substrate for the labelling enzyme is added with the perfusing buffer. On introduction of the enzyme-labelled antigen together with the native antigen, the enzyme reaction started. When the excess enzyme reached the detector region of the system, a peak with very high product concentration was seen. However, after the peak concentration was washed out, the descending part of the curve represented the conversion catalyzed by the affinity bound enzyme in the immunosorbent. By eliminating the separate washing step and the separate addition of substrate, it was possible to reduce the time for one assay cycle from 6–7 min to 70 s [17].

Whether this is an achievement that is useful or not, only the future will show. At present, not enough is known about the proper frequency of analysis. In many cases once every 15 minutes is appropriate for monitoring established processes.

The fully computor-controlled flow-injection ELISA is carried out with the equipment illustrated in Fig. 5. There is very little manual handling, so it is possible to reduce variations between assays to an absolute minimum. In the assay cycles, however, the dissociation step involves treatment of the immobilized antibodies under harsh conditions, which leads to a slow but significant denaturation. This occurs when running an assay repeatedly (Fig. 3). As stated above, the competitive situation between native and labelled antigen is constant, it is only the amplitude of the response that changes. The capacity of the column is determined intermittently by adding a sample containing only labelled antigen. The capacity value obtained is used in the evaluation process. Thus, by expressing the calibration curve in terms of an equation, it is possible to adjust the evaluation intermittently

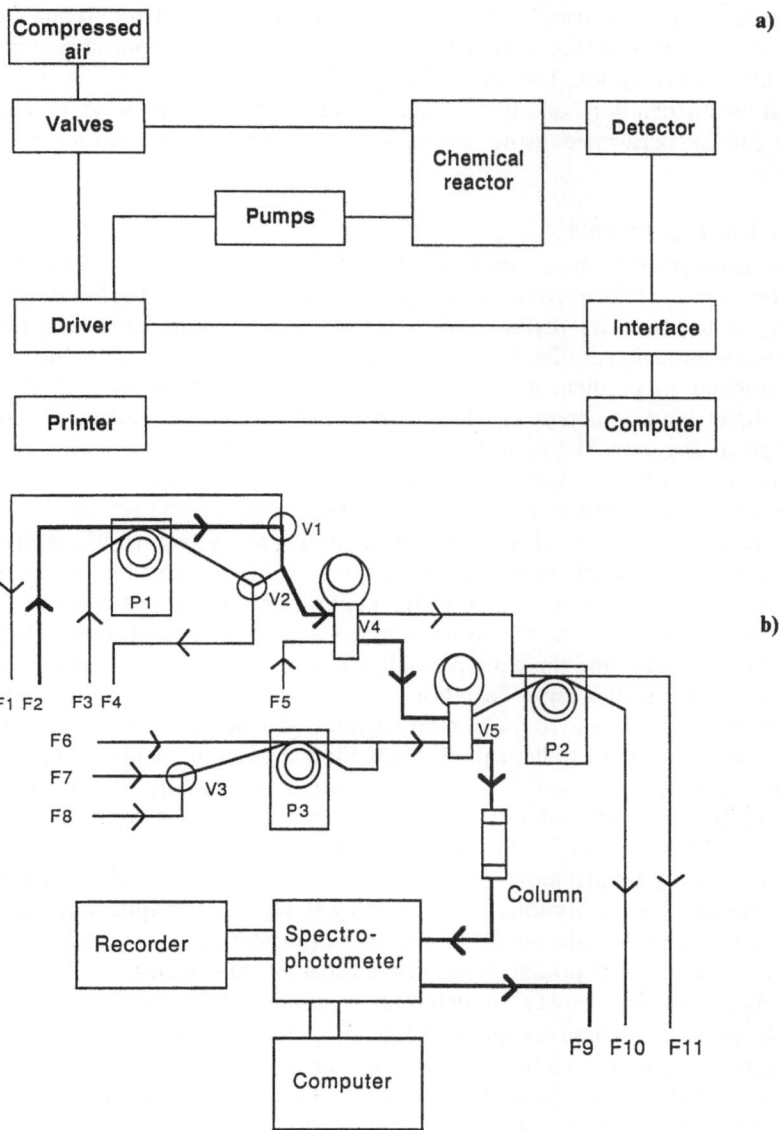

Fig. 5a. Block diagram illustrating the arrangements with the computorized flow injection ELISA. **b)** Schematic presentation of the experimental set-up in the computorized flow injection ELISA. The different symbols in the figure represent the following: *F1*: excess of main buffer; *F2*: main buffer; *F3*: washing buffer; *F4*: excess flow of washing buffer; *F5*: substrate solution; *F6*: solution with enzyme labelled antigen; *F7*: sample solution; *F8*: reference solution; *F9–11*: waste; *P1–P3*: peristaltic pumps; *V1–V3*: three-way valves; *V4* and *V5*: injection valves. (From Ref. 18, with permission)

by compensating for changes in the capacity of the immunosorbent. When the decrease in capacity is compensated for in this manner, the response remains constant over many cycles. The stability depends on the strength of the interaction and the dissociation conditions needed, but in many cases several hundreds of assays can be performed using the same packed bed of immobilized antibodies.

2.3.2 Gradient Experiments

Besides traditional calibration curves it is of importance to known how well the analytical system can follow dynamic changes in a reactor. One way to characterize the system is to study its performance when it is exposed to increasing and decreasing concentration gradients of a metabolite. We have tested the flow-binding assay equipment in gradient experiments, both with the competitive ELISAs and with direct binding assays in which enzymes are the target molecules. The results of such assays are shown in Fig. 6. It is quite clear from these figures that the flow-binding assay is efficient in registering dynamic changes. Furthermore, it is clear that the competitive binding assay is more accurate and reliable in the lower concentration regions, whereas the direct binding assay gives good readings at higher concentrations as well. This performance of the competitive assays is expected when considering the shape of the calibration curves. The accuracy is highest in the steepest part of the curve, whereas at higher concentrations there are difficulties in evaluating the data. A small deviation in the signal registered in this region will result in a large deviation in the concentration value read. The easiest way to handle this is to make a larger dilution when the reading reaches a certain level by using a smaller sample loop. Since FIA equipment is very well suited for arranging reproducible and exact dilutions of samples, this is not a major task. However, the concept of varying the degree of sample dilution depending on the concentration has not yet been practically implemented.

Flow-ELISA may be attractive for monitoring cultivations as well as downstream processes. In fermentations it is interesting to register the appearance and concentration of certain substances. In the area of downstream processing it is desirable to trace a target protein through the different unit operations. Usually only pH, A_{280} and conductivity are determined on-line, whereas many analyses are carried out off-line. These are time and labour intensive and the results usually appear when it is too late to apply the analytical result to control the process. In the light of the fact that 40–80% of the total cost for a biotechnological process lies in the downstream steps, it is amazing that so little has been done to better monitor these processes. In general, very few instruments are available on the market to monitor downstream processes, even if only the three parameters mentioned above are included [21].

A chromatographic procedure is often regarded as too quick to be suitable for process monitoring. However, a totally different situation applies in large scale operation as compared to the analytical procedures used in many academic laboratories.

In a series of experiments the automated equipment was used for tracing and quantifying certain proteins when mixtures of proteins were separated

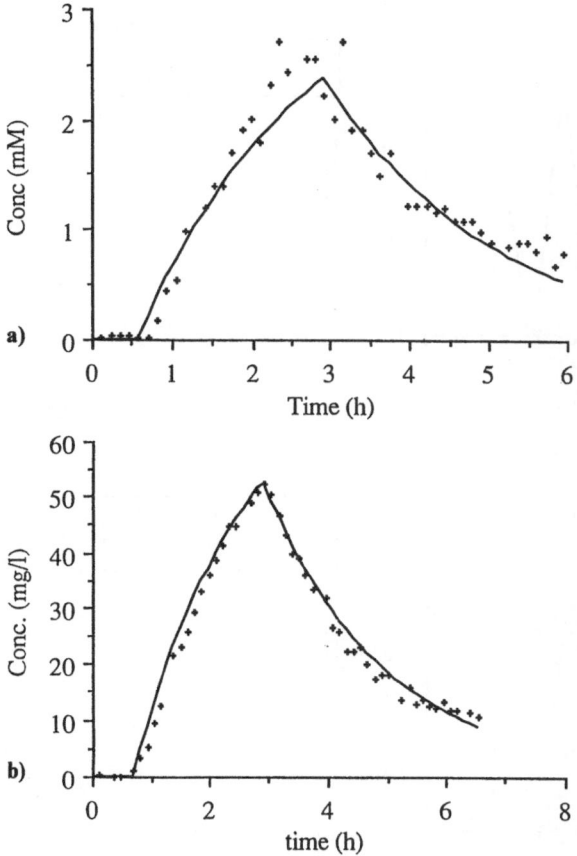

Fig. 6a, b. Measurements of increasing and decreasing gradients using flow injection binding assays. The system studied was concanavalin A bound to the affinity support. Horseradish peroxidase and α-methyl-D-gluconopyranoside were allowed to compete in the binding reaction. The *solid lines* represent the theoretical gradients and the *crosses* symbolize the data points. **a)** Results from a competitive binding assay. **b)** Results from a direct binding assay

by chromatography. Figure 7 shows three chromatograms obtained with the same equipment, but with different affinity columns and different enzyme labelled antigens. The three compounds quantified were IgG, horseradish per-oxidase, and human serum albumin. IgG and albumin were monitored in a competitive assay. The horseradish peroxidase was included as an example of a direct binding assay in which the catalytic actitivity of the target molecule can be utilized directly in the quantification step. It is more convenient to use direct binding assays at high concentrations of the target molecule, provided it can be quantified either as shown here, or by an enzyme-labelled second antibody [18, 19, 22].

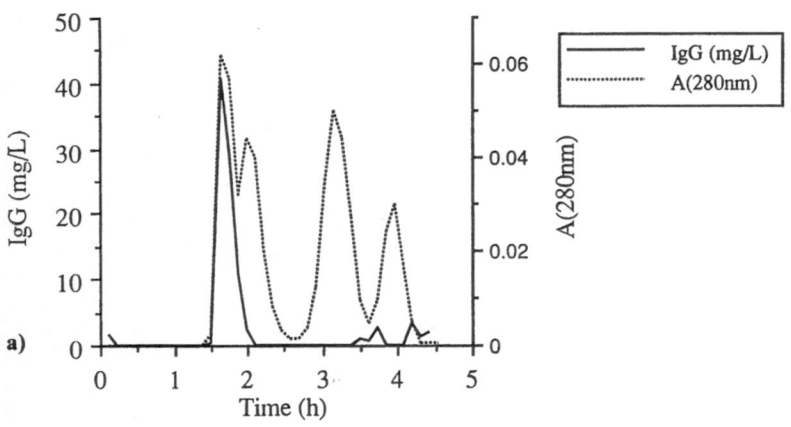

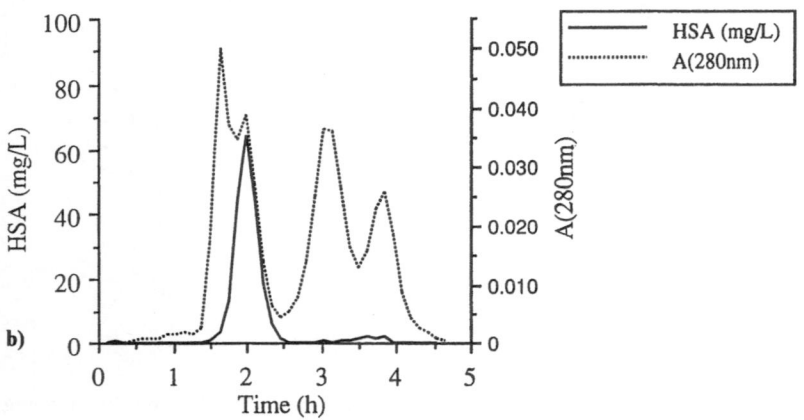

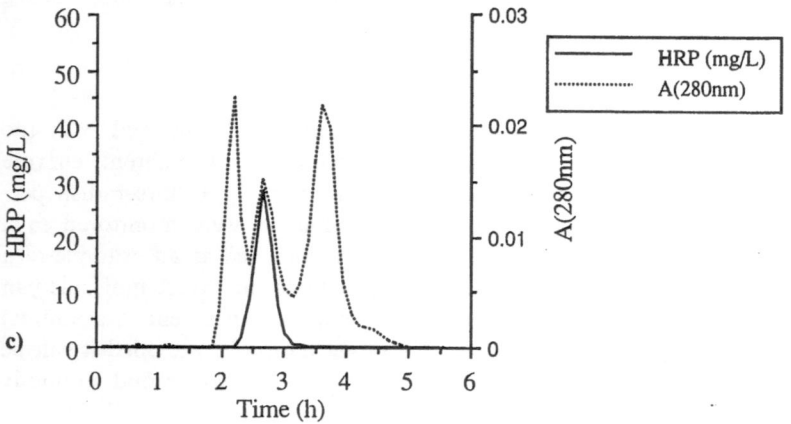

2.4 Other Flow-Injection Immunoassays

In addition to using traditional binding reactions, it is also possible to utilize turbidimetric assays. The aggregate formation between polyvalent antigens and two or more antibodies is utilized for the quantification. The aggregates formed scatter light and thus it is possible to quantify the degree of aggregate formation. The immunocomplexes formed are quantified using a nephelometer. The technique is very versatile, but consumes a lot of reagents since large amounts of antisera are used. Furthermore, under conventional conditions such assays will operate at lower sensitivity than regular immunochemical binding assays. It may, however, be fully sufficient in many cases. One great advantage of such an approach is that it is possible to operate in a continuous manner. The stream from the reactor containing the molecule to be quantified is mixed with a stream of antisera. After proper mixing and reaction in the flow system, measurement is performed using a flow cell. As in most of these assays, the first published example of an FIA-turbidimetric assay appeared in clinical chemistry. In efforts to quantify human IgG, the IgG was mixed with goat antiserum specific against human IgG. To further improve the binding reaction polyethyelene glycol and sodium chloride were added to the reaction mixture [23].

In recent publications it was also shown that it is possible to utilize an immunoprecipitation reaction to quantify monoclonal antibodies and pullulanase [24, 25].

3 Direct Monitoring of the Binding Reactions

Direct monitoring of the antigen-antibody interaction is very attractive since it would make it possible to design analytical systems without the need for additional reagents except the immobilized immunochemical. In situ monitoring would then be possible. This field has not developed far. The present stage of development focuses on the design of new methods to register the binding reaction.

A whole range of different techniques has been used to register direct binding interactions. Some of these are listed in Table 3.

◀ **Fig. 7a–c.** Evaluation of Flow-injection ELISA to monitor column chromatography. A mixture of proteins were chromatographed through a Sephadex G-200 column and the effluent was analyzed by A_{280} and by sampling using the computorized flow-ELISA equipment. **a)** Detection of IgG in the separation of IgG, human serum albumin, lysozyme and acetyl-tryptophane. **b)** Detection of human serum albumin in the same chromatographic separation. **c)** Detection of horseradish peroxidase when separating a mixture of IgG, peroxidase and lysozyme by chromatography

Table 3. Examples of techniques utilized to register direct binding reactions when developing immunobased biosensors

Technique	Ref.
Electrochemical methods	
Streaming potential	[44]
	[28]
	[44]
Potentiometric electrodes	[27]
Piezoelectric crystals	[46]
Optimal methods	
Surface plasmon resonance	[47]
Reflectometry	[48]
Ellipsometry	[32]

3.1 Electrochemical Methods

The basis for these measurements is the difference in charge initiated by binding. Either the bound substance carries many charges, thereby introducing a change that can be registered, or the binding per se leads to an exposure of a different number of charges to the electrochemical transducer. Still another alternative is that binding per se initiates conformational changes leading to a physico-chemical change that can be registered by the transducer.

In the area of potentiometric immunosensors most work has been carried out using lectin-carbohydrate interactions as a model system. Janata [26] coupled concanavalin A to a thin polyvinyl chloride layer on a platinum wire. By this arrangement they were able to detect carbohydrate binding to the lectin. The binding between membrane bound antigen and free antibody led to a potential shift that was utilized by Aizawa et al. [27]. There is still room for more research before these sensors are ready to leave the research laboratories and to be implemented in real measuring situations.

3.1.1 Streaming Potential

Streaming potential is another approach to electrochemical detection of the direct binding reaction between antigen and antibody. In this case there are also some technical limitations that have to be overcome before the technique can be more generally applicable. The basis for streaming potential measurement is the fact that if a solid surface comes in contact with a solution, there is an electrochemical potential built up between the bulk solution and the surface. Such systems have long been utilized when characterizing various inorganic materials. When a solid support is modified to allow coupling of affinity ligands, it becomes possible to register the streaming potential prior to, during and after exposure of this surface to a compound capable of binding the ligand. The potential read from a streaming potential measurement is dependent on many factors, including the flow rate of

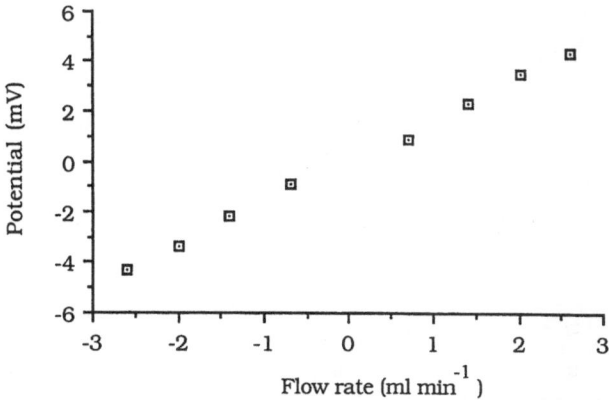

Fig. 8. Streaming potential measured over a bed (diameter 5 mm) of packed Octyl-Sepharose at different flow rates of Tris-HCl buffer pH 7.6

the liquid passing the solid support. When the support is packed in a column and the streaming potential is read between electrodes at the top and the bottom of the column, it is possible to determine the potential dependence upon flow rate. Figure 8 shows the results of pumping buffer over a column of Octyl-Sepharose. A characteristic of streaming potential is that when the flow is reversed, the potential changes sign. This also means that at a flow rate of zero, there is no streaming potential. In Fig. 8, this can be clearly seen.

On addition of bovine serum albumin to the Octyl-Sepharose, the albumin binds and in theory should induce a change in the charge distribution. This would be recognized as a shift in streaming potential. Figure 9 demonstrates that this is exactly what happens [28].

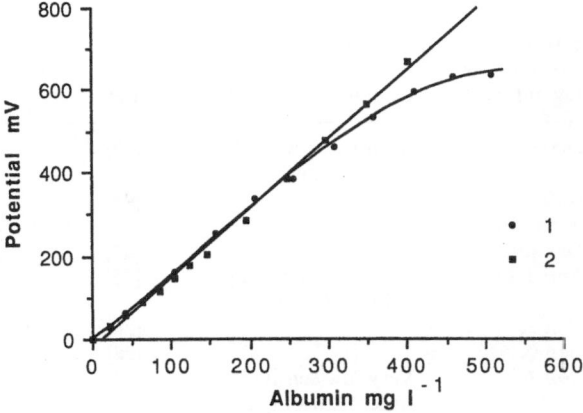

Fig. 9. Streaming potential read at varying concentrations of albumin injected into the flow system. The column was filled with Octyl-Sepharose (from Ref. 28, with permission)

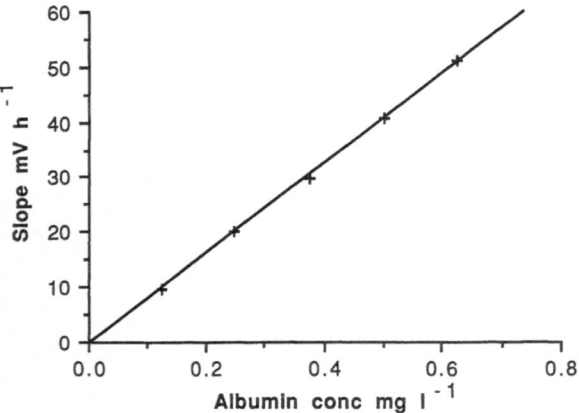

Fig. 10. Streaming potential read during continuous measurements as the slope of the plot potential vs time at different concentrations of albumin vs the albumin concentration fed at the moment the potential was read

It is noteworthy that the target molecule, in this case albumin, can be continuously fed to the affinity sorbent and a continuous change in the potential can thus be monitored (Fig. 10). This is possible as long as there is no saturation effect registered. By varying the concentration of albumin in the feed stream, it was possible to demonstrate that the slope of the curve when recording the streaming potential vs time was directly proportional to the concentration of albumin in the feed. This observation means that the first derivative of the signal at any moment can be used as a reading of the actual concentration of target molecule passing the adsorption column at that moment, i.e. a true continously analysis.

Streaming potential has been used to study many different interactions. Some of these are listed in Table 4.

Table 4. Affinity pairs studied using streaming potential

Molecule measured	Ligand used	Comments	Ref.
Human serum albumin	Octyl-	Both pulse-wise and continuous	[28]
Transferrin	Anti-transferrin		[44]
Concanavalin A	Dextran		[44]
mcAb	Anti-IgG	Process monitoring hybridoma culture	[45]
Horseradish peroxidase	Con A		[28]
Invertase	Con A		[28]
Dextran	Con A	Very low sensitivity, since dextran has very few changes	[28]
IgG	Protein A		[44]

From the table, it can be seen that many different interactions may be studied and quite a range of target molecules can be quantified. Before streaming potential becomes a widely used analytical technique for monitoring affinity interactions stable and reproducible electrodes must be developed. A series of different electrodes has been tested, but no adequately stable system has been found. Therefore, streaming potential seems very promising, but in its present state of development it is only suitable for use in research laboratories devoted to the development of analytical instrumentation.

4 Optically Based Immunosensors

There are many advantages of optical detection systems. They are isolated electrically from the measuring site, which is of vital importance for in vivo measurements. Furthermore, they are extremely resistant to electrical disturbances. Optical systems often have very short response times and a wide range of materials can be used for the fabrication of the sensors for these systems. Another advantage of optical sensing is that if fouling occurs on the membrane surface of the sensor, the amplitude of the signal from the sensor will be unaffected and only the response time will be prolonged. This is in contrast to other systems where the amplitude decreases with increased fouling.

The main drawbacks to optical sensor systems are that they are fragile, rather bulky and expensive. However, in recent years the development of semiconductor lasers, fibre optics and micro-mechanics have made it possible to overcome many of the obstacles of the optical systems.

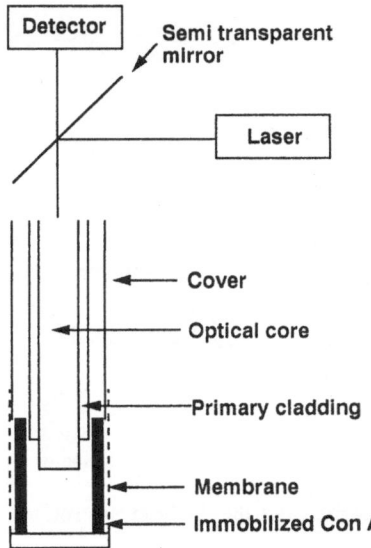

Fig. 11. Basic principle of a needle optic sensor based on fluorescein-labelled dextran

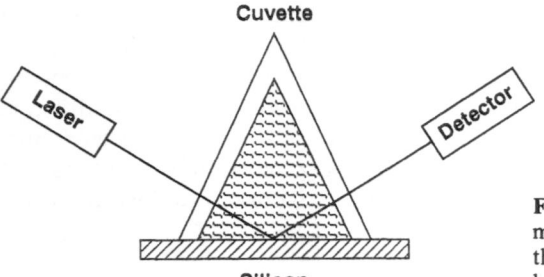

Fig. 12. The principle of a reflecto-meter. The cuvette can be of flow-through type and the substrate can be made of silicon

Various optical techniques have been evaluated for application to immunosen-sing such as adsorption techniques, fluorescence techniques [29, 30], reflection techniques [31], ellipsometry [32], surface plasmon resonance [33] and evanescent field optics [34].

An example of the fluorescence technique is the glucose sensor system described by Schultz and Mansouri [29], see Fig. 11. They use fiber optics in combination with a permeable membrane and fluorescence-labelled-dextran affinity bound to immobilized concanavalin A at the inner surface of the membrane. The principle of the system is that glucose molecules penetrate the membrane through the pores. This causes some labelled dextran to be displaced from the concanavalin A binding sites and thereby to enter the light-path of the exciting light which is led through the fiber. The fluorescent light emitted, which has a considerably longer wavelength, is detected through the same fiber. The amount of fluorescent light detected by the photomultiplier detector increases as the concentration of glucose increases. This system has now been miniaturized to the size a hypodermic needle. Detection can be made down to less than 1 g l^{-1}, which is sufficient for use in patient monitoring.

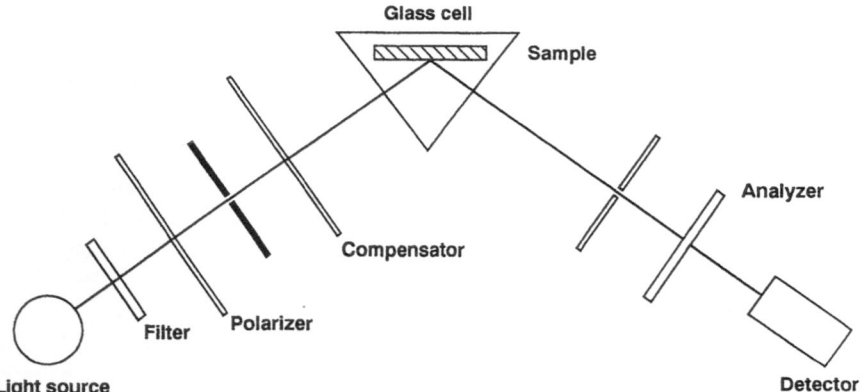

Fig. 13. An example of an ellipsometer setup. The polarizer and the analyzer are rotated until the light to the detector is extinguished

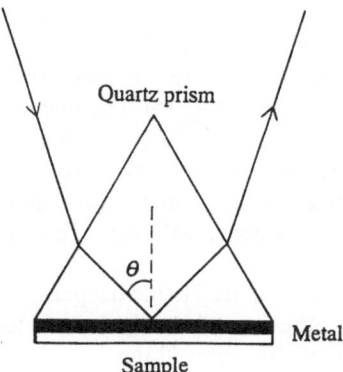

Quartz prism

θ

Metal

Sample

Fig. 14. The basic system for surface plasmon resonance measurements

Reflectometry is based on the principle of detecting changes in the intensity of reflected light [31]. The sensitivity of such systems can be increased by using polarized light, which is parallel to the plane of incidence at a certain angle (the Brewster angle). The basic principle is shown in Fig. 12. Arwin and Lundström [31] have used this system for determination of tryptic digestion of human immunoglobulin G (IgG).

In ellipsometry the change in the degree of polarization of light is detected. This technique is more complex than reflectometry. An ellipsometer consists of a light source, a polarizing filter, a test surface, an analyzing component and a detector (Fig. 13). Jönsson et al. [32] have used this method for determination of immunoglobulin G and protein A down to concentrations of 1 mg ml^{-1}.

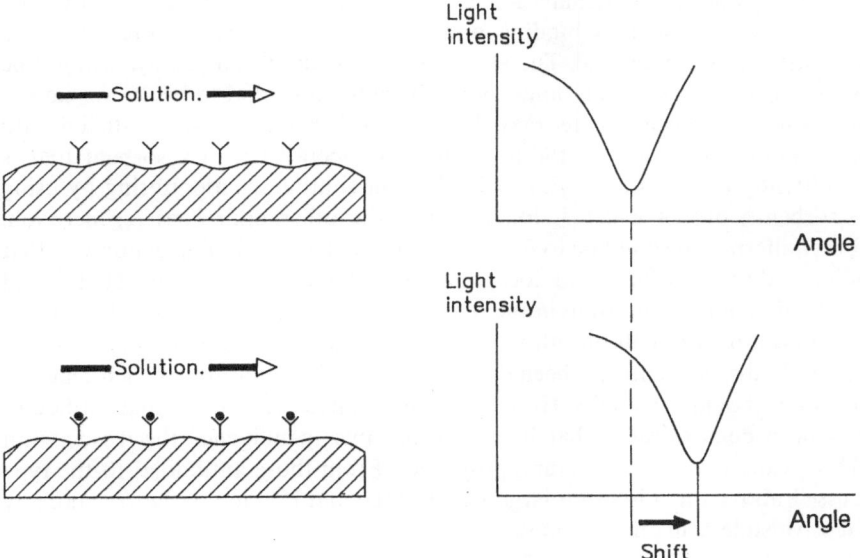

Fig. 15. Surface plasmon resonance causes an angle shift when the ligands are coupled

The principle utilized in surface plasmon resonance is the effect of collective oscillation of the conducting electrons at the surface of a metal. The basic setup of a surface plasmon resonance system is shown in Fig. 14. If the incident angle varies, the output will show a minimum for a certain angle (Fig. 15). The medium outside the metal film and any adsorbate on the metal surface will cause a shift in the angle of the minimum. This technique has been used for antigen-antibody reaction of surfaces [34]. The sensitivity is better than that of ellipsometry and more applications in the biological field based on plasma resonance will surely turn up in the future.

Sensors which are based on evanescent field effects do not need to have separation or washing steps to separate the bound label from the free label. The reason for this is that the unbound label is not monitored by the detector [34].

Optical systems are still rather complicated but in the near future the new optoelectronic semiconductor-based components such as lasers and detectors will make this technique much easier to use and also much less expensive.

5 Selection of Antibodies

Continuous immunological monitoring in general raises some very special demands on the participating reagents. In traditional immunoassays, the most efficient and strongest binders possible are used. The better the binding constant, the better the sensitivity in the assay. When dealing with immunosensors, some other criteria come into play in the selection of proper antibodies. In repeated binding assays, e.g. flow-ELISAs, the antibodies must be efficient in binding, but they also must be efficient in releasing the bound material under reasonable conditions under which the antibody will remain active. This means that in this case, antibodies which are less efficient in binding and with good properties with respect to dissociation must be selected. Therefore, these non-equilibrium assays will not be as sensitive as the more traditional assays have the potential to be. In the case of continuous monitoring there may be some other criteria to be taken into consideration. For example, the assay may be operated with a sorbent that is intermittently regenerated. This can be done in continuous monitoring by using two sorbents, one of which is loaded while the other undergoes regeneration. Another alternative would be to find antibodies with such binding conditions that binding and dissociation are almost equal under the conditions used. That would lead to a dynamic equilibrium in the sense that the amount bound on the sorbent would reflect the free concentration in the surrounding medium at every moment. Such antibodies have not yet been described, and it is doubtful that such reagents would offer enough sensitivity. However, with the introduction of switch antibodies it has been demonstrated that it is fully possible to influence the dissociation without really ruining the binding properties. From those antibodies to the type discussed above there is still a large step to take and it is too early to make any conclusive statements in this matter.

One property that has not been ranked as important in selection of polyclonal antibodies is stability. It is now obvious that the stability properties of the individual clones of monoclonal antibodies differ quite a lot. Most selection work has been performed on the basis of binding constants, but with the applications discussed in this chapter in mind it may be more important to screen for stable antibodies with moderate binding constants.

6 Concluding Remarks

The development of immunobased biosensors still lags behind the development of enzyme based sensors by several years. This could be remedied if the common problems were addressed by those concerned with developing the enzyme-based sensors, and this would also mean that a more rapid implementation of immunosensors would be possible. There is, however, a fundamental problem to be solved: how will all the new information generated by this totally new generation of sensors be utilized? At present it is not known what the proper analytical frequency, time constants, etc. may be necessary for monitoring a given process. It may be that the first applications will deal with this expansion of knowledge and only later will this knowledge be utilized in the application of immunobased sensors to process control of bioconversions and downstream processing steps.

The area has been neglected for many years, but we can now see a lot of interesting development concerning new sensor systems, new approaches to improve operational stability of the immunochemicals and also a willingness to apply such sensors.

Acknowledgements: This project was supported by The National Swedish Board for Industrial and Technical Development (NUTEC) and The Nordic Fund for Technology and Industrial Development.

7 References

1. Clark LC Jr, Lyons C (1962) Ann NY Acad Sci 102: 29
2. Turner APF, Karube I, Wilson GS (eds) (1987) Biosensors. Fundamentals and Applications. Oxford Univ Press, Oxford, UK
3. Danielsson B, Mattiasson B, Karlsson R, Winqvist F (1979) Biotechnol Bioeng 21: 1749
4. Schügerl K (1988) Anal Chim Acta 213: 1
5. Holst O, Håkanson H, Miyabayashi A, Mattiasson B (1988) Appl Microbiol Biotechnol 28: 32
6. Lin KH, Iijima S, Shimizu K, Hishinuma F, Kobayashi T (1989) Appl Microbiol Biotechnol 32: 313
7. Scheper T, Brandes W, Grau C, Hundeck HG, Reinhardt B, Rüther F, Plötz F, Schelp C, Schügerl K, Schneider KH, Giffhorn F, Rehr B, Sahm H (1991) Anal Chim Acta 249: 25
8. Mattiasson B (1984) Trends Anal Chem 3: 245
9. Rubinstein K, Schneider R, Ullman E (1972) Biochem Biophys Res Commun 47: 846
10. Chang JJ, Crowl CP, Schneider RS (1975) Clin Chem (Winston-Salem) 21: 967
11. Schneider RS, Lindqvist P, Wong RC, Rubinstein KE, Ullman EF (1973) Clin Chem (Winston-Salem) 19: 821
12. Burd JF, Wong CR, Feeney JE, Carrico RJ, Boguslaski RC (1977) Clin Chem (Winston-Salem) 23: 1402
13. Mattiasson B (1980) J Immunol Methods 35: 137
14. Mattiasson B, Ling TGI (1980) J Immunol Methods 38: 217

15. Mattiasson B, Borrebaeck C, Sanfridsson B, Mosbach K (1977) Biochim Biophys Acta 483: 221
16. Mattiasson B, Svensson K, Borrebaeck C, Jonsson S, Kronvall G (1978) Clin Chem 24: 1770
17. Mattiasson B, Berdén P, Ling TGI (1989) Anal Biochem 181: 379
18. Nilsson M, Håkanson H, Mattiasson B (1991) Anal Chim Acta 249: 163
19. Nilsson M, Håkanson H, Mattiasson B (1991) Submitted for publication
20. Chung S, Wen X, Vilholm K, De Bang M, Christian G, Ruzicka J (1991) Anal Chim Acta 249: 77
21. Mattiasson B, Håkanson H (1991) Measurement and control in downstream processing. In: Carr-Brion K (ed) Measurement and control in bioprocessing. Elsevier, Amsterdam
22. de Alwis U, Wilson GS (1987) Anal Chem 59: 2786
23. Worsfold PJ, Hughes A & Mowthrope DJ (1985) Analyst 110: 1303
24. Freitag R, Scheper T, Spreinat A, Antranikian G (1991) Appl Microbiol Biotechnol 35: 471
25. Freitag R, Fenge C, Scheper T, Schügerl K (1991) Anal Chim Acta 249: 113
26. Janata J (1975) J Am Chem Soc 97: 2914
27. Aizawa M, Kato S, Suzuki S (1977) J Membr Sci 2: 125
28. Mattiasson B, Miyabayashi A (1988) Anal Chim Acta 213: 79
29. Schultz JS, Mansouri S (1988) Optical fiber affinity sensors. In: Mosbach K (ed) Methods in enzymology, vol 137. Academic, New York, p 349–366
30. Schultz JS (1991) Scientific American August 40: 48
31. Arwin H, Lundström I (1988) Surface-oriented optical methods for biochemical analysis. In: Mosbach K (ed) Methods in enzymology, vol 137. Academic, New York, p 366–381
32. Jönsson U, Malmqvist M, Rönnberg I (1985) J Colloid Interface Sci 103: 360
33. Sutherland RM, Dahne C, Place JF, Ringrose AS (1984) Clin Chem 30: 1533
34. Robinson GA (1989) Optical Immunosensors. Biosensors 89. Cambridge, UK
35. Mattiasson B, Larsson KM (1987) Flow-injection enzyme immunoassay – a quick and convenient binding assay. In: Neissel OM, van der Meer RR, Luyben KCAM (eds) Proc. 4th European Congress Biotechnol, vol 4. Elsevier, Amsterdam, p 517
36. Alwis WU de, Wilson GS (1985) Anal Chem 57: 2754
37. Heineman WR, Halsall HB (1987) Enzyme immunoassay with electrochemical detection in flow systems. In: Schmid RD (ed) Biosensor International Workshop 1987 (GBF monographs 10), VCH Verlagsgesellschaft, Weinheim, Germany, p 127
38. Hughes A, Worsfold PJ (1985) Anal Proc, 22: 16
39. Kelly TA, Christian GD (1982) Talanta 29: 1109
40. Mattiasson B, Nilsson H (1977) FEBS Lett. 78: 251
41. Birnbaum S, Bülow L, Hardy K, Danielsson B, Mosbach K (1986) Anal Biochem 158: 12
42. Lee IH, Meyerhoff ME (1988) Mikrochim Acta 111: 207
43. Maeda M, Tsuji A (1985) Anal Chim Acta 167: 241
44. Glad C, Sjödin K, Mattiasson B (1986) Biosensors 2: 89
45. Miyabayashi A, Mattiasson B (1990) Anal Biochem 184: 165
46. Roederer JE, Bastiaans GJ (1983) Anal Chem 55: 2333
47. Liedberg B, Nylander C, Lundström I (1983) Sensors Actuators 4: 299
48. Kronick MN, Little WA (1975) J. Immunol. Methods. 8: 235

A Human Genome Analysis System (HUGA-I) Developed in Japan

I. Endo
Chemical Engineering Laboratory, The Institute of Physical and Chemical Research, RIKEN, Wako-shi, Saitama 351-01, Japan

A part of the Japanese human genome project has resulted in the development of a world original human genome analysis system. Circumstances of the project, objective, design concept, and constitution of the system are described in the paper.

1 Introduction

Since 1988, the human genome analysis project, entitled "Studies on Gene Structure", has been promoted by the Japanese Science and Technology Agency (STA). By developing this project, Japanese scientists and engineers hope to play their roles in the world not only in the elucidation of human biology and medicine but also in the development of basic biology.

The aims of this project are:

1) To develop an automated rapid analytical system for sequencing human genome DNA; and
2) To elucidate preparation techniques for the genome sequences, including preparation of genome libraries, mapping techniques and database systems.

Scientists and engineers from 12 research facilities including universities, government institutions and private companies have participated in this project and research costs of about 600 million Yen ($ 4.5 million) have already been funded by STA. Work on the first objective of the project has been led by the author and that on the second one has been led by Professor Yohji Ikawa.

The process of determining the nucleotide sequence of large DNA molecules, like human genes, is tedious and time consuming, yet requires our untiring energy. It would be a significant contribution to the development of medical science and biology if we could solve these problems by automating, accelerating and systematizing the conventional process. This idea was first proposed by Professor

Advances in Biochemical Engineering
Biotechnology, Vol. 46
Managing Editor: A. Fiechter
© Springer-Verlag Berlin Heidelberg 1992

Wada [1]. He and his team of scientists and engineers have developed several automated machines with the assistance of the Special Coordination Fund for the Promotion of Science and Technology (1981–1984) and the Cancer Research Fund (1985–1987) under the auspices of the STA. These funds allowed for the completion of machines like the DNA extractor (Seiko Instr.), dideoxy reactor (Seiko Instr.), gel-forming machine (Fuji Film Co.) and the fluorescent sequencer (Hitachi). They are no available world-wide.

The authors have already introduced the essentials of a world original human genome analysis system (HUGA-I) in Nature [2]. The objective of this article is to describe the system in more detail by using photographs and figures.

2 System Components

Our prototype model, HUGA-I, is designed to accelerate the functions of the machines already developed and to combine the whole sequencing process according to the shotgun process which was proposed by Deininger [3]. It consists of the following main unit processes: construction of the human DNA libraries, cloning of DNA fragments, purification of these fragments, dideoxyreactions, injection of DNA molecules from the dideoxy reactor into the gel, sequencing gel electrophoresis, and editing numbers of short sequences into the original DNA sequence.

Such processes as the construction of the DNA libraries, sonication of the template DNA (~ 100 kb) into smaller sizes (~ 1 kb), cloning and multiplying the smaller DNA fragments by using M13 phages and *E. coli* cells could not be automated because these processes are unpredictable. However, the plaque selection from the plate and transferring of the microorganisms into the test tubes are routine operations. Seiko Instrument Inc. has accomplished this operation using a robot arm as shown in Plate 1. The downstream processes, from

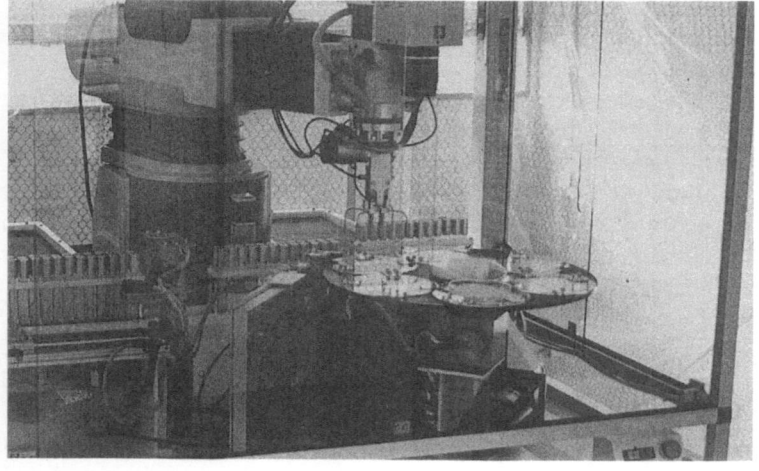

Plate 1. Plaque Selector (Seiko Instrument Inc.)

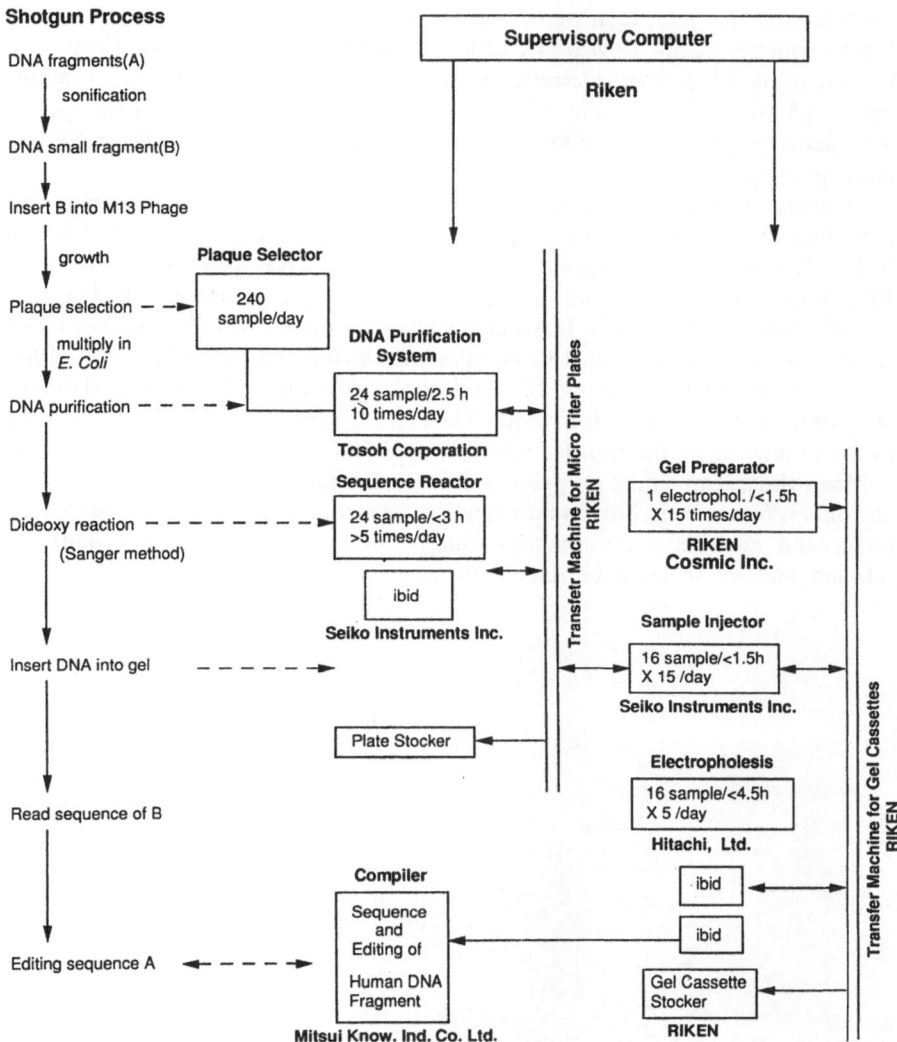

Fig. 1. Block diagram of the Human Genome Analysis System (HUGA-I)

the purification of DNA fragments to editing the DNA sequences, have been completely automated by introducing two kinds of transfer robots; one for the transfer of the micro titer plates (MTP, hereafter) and the other for the transfer of the gel cassette. A schematic diagram of the system is shown in Fig. 1.

3 Concept of HUGA-I

The concept of our system is as follows: If we can elucidate 100 kb of human DNA sequences per day with the system and could operate it for 300 days continuously per year, then we could read the 3 billion base pairs of the human DNA sequence within 10 years by using 10 lines of this system. The Hitachi Co. has already developed a fluorescent sequencer which can read 450 bases per sample within 4.5 hours. The capacity of the slab gel is 16 samples, so that if we operate 3 sequencers simultaneously for 5 cycles in a day, then we can analyze 108,000 bases per day.

In order to operate Hitachi's sequencer efficiently, operating schedules and capacities of upstream processes were determined according to the flow diagram in Fig. 1. RIKEN and Cosmic Inc. have made a gel preparation machine (Gel Preparator) in which the slab gel (polyacrylamide gel, 0.3 mm in thickness) is solidified and cleaned. Seiko Instruments has developed a sample injection robot which can automatically inject 16 samples into the top of a slab gel cassette within 1.5 hours (Plate 2). After all of the samples are injected into the slab gel, electrophoresis takes place for 1 hour. The cassette (about 8 kg) is then transferred to the sequencer by the transfer robot which RIKEN has developed (Plate 3).

Meanwhile, samples are prepared as follows: human DNA cloned into a cosmid vector is sonicated and broken into smaller fragments, and then cloned using M13 phage and *E. coli* systems. Plaques which contain human DNA fragments are selected and innoculated automatically into test tubes by the plaque selector

Plate 2. Sample Injector (Seiko Instrument Inc.)

Plate 3. Gel Cassette Transfer Robot (RIKEN)

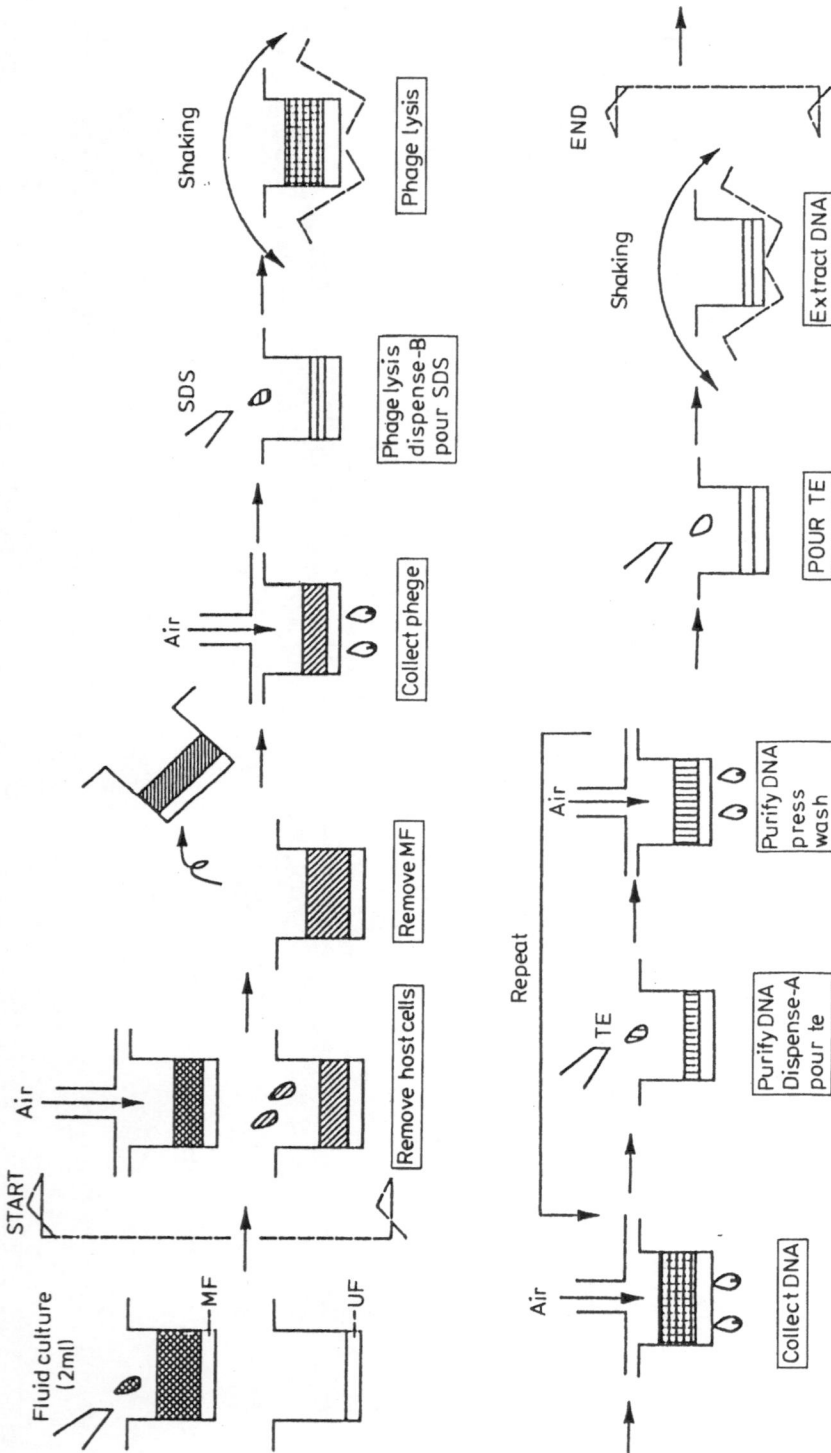

Fig. 2. Extraction and purification of the template DNA from M13 culture fluid, using the differential filtration method

(Plate 1). This machine provides us with 240 samples per day. The DNA fragments then multiply in test tubes for 6 hours and are purified by a DNA purification system developed by the Tosoh Corporation. Tosoh has developed a membrane method in order to accelerate the purification operation. The DNA fragments, *E. coli* cells and phages are successively filtered first by an MF membrane and then by a UF membrane. The DNA thus obtained are redissolved in a Tris-EDTA solution, and 1 µl of this solution is pipetted automatically into 24 designated wells in the MTP (94 wells). This operation is executed by the system within 2.5 hours, so we can operate this machine at least 9 times per day. The extraction and purification of DNA fragments from M13 culture fluid are shown in Fig. 2.

After the pipetting operation is over, another transfer machine developed at RIKEN picks up the MTP and transfer it to the dideoxy reactor made by Seiko Instruments [4]. This reactor, also called a chemical robot can execute faithfully and automatically a complete set of Sanger's sequencing reactions. The major functions achieved by this machine are quantitative addition and dispensing of reagents, mixing, heating, and cooling all within the unit. Using this machine, we can accelerate the dideoxy reaction for 24 samples to within 3 hours and operate it more than 5 times per day. In order to provide DNA molecules to the sequencers, we have installed 2 sets of these machines. After the Sanger's reaction is over, the transfer machine picks up the MTP again and hands it to the sample injector which was mentioned earlier.

4 Determination of DNA Sequences

RIKEN has made a computer program for sequence assembly and editing in collaboration with the Mitsui Knowledge Industry Co. In the assembly of random sequence fragments, the computer links up matching portions of the fragments to reconstruct the complete original sequence. Where ambiguous data exist, the computer leaves in the sequence. To increase the accuracy of this approach, fragments which contain the same portions of the original sequence are automatically read from different directions and different locations approximately four times. It should be noted that inevitable errors occurring in the raw sequence data will be corrected or reduced by overlapping these sequences. An improved program for compilation of the sequence is being prepared by integrating our experiences. It will be operational on any VAX computer.

In this fiscal year (1991–1992), we will evaluate and examine the performance of HUGA-I by using DNA cosmids and we will begin the analysis of the human genome starting in the next fiscal year. Even though HUGA-I has the goal of elucidating the human genome, it is by no means limited to this. It can be used to help map out the genetic sequence of any living organism. The author would also like to point out that the development of the hardware must go hand in hand with the methodology in order that additional significant progress can be made. In this fashion, such time consuming and monotonous work can be taken away from the scientist so that new original research can be conducted.

Acknowledgements: The author would like to express his heartful thanks to Dr. E. Soeda, Dr. Y. Murakami, Mr. K. Nishi, the STA, the contributing companies and Professor A. Wada for their substantial contributions to this project.

5 References

1. Wada A (1987) Nature 325: 771
2. Endo I et al (1991) Nature 352: 89
3. Deininger PL (1983) Anal Biochem 129: 216
4. Sakabe M et al (1990) Rev Sci Instrum 61: 1966

Optimization and Control in Fed-Batch Bioreactors

S. Shioya

Department of Biotechnology, Osaka University, Suita, Osaka 565, Japan

A method for obtaining maximum production of the bioproduct in fed-batch cultures is explained, and its validity is demonstrated by experimental data. The approach is based on a model which describes the relationship between the specific production rate, ϱ. and specific growth rate, μ. Using a mathematical model, an optimal profile of the specific growth rate could then be obtained easily by the Maximum Principle. Finally, the optimal profile was realized by changing the feed rate of the substrate in a practical fed-batch culture. Practical examples of bioproduction, such as histidine, lysine, and glutathione, as well as production of an enzyme, showed that the two-stage production process could be realized experimentally, thus demonstrating the validity of the method.

Advances in Biochemical Engineering
Biotechnology, Vol. 46
Managing Editor: A. Fiechter
© Springer-Verlag Berlin Heidelberg 1992

List of Symbols

a_1	model parameter in Eq. (7) (i = 1, 2)
b_1	model parameter in Eq. (7) (i = 1, 2)
E	enzyme content of cells ($U\,g^{-1}$)
F	feed rate of substrate ($l\,h^{-1}$)
$\bar{F}$	nominal feed rate of PF system ($l\,h^{-1}$)
ΔF	feed rate for compensation of disturbance ($l\,h^{-1}$)
F_{NH3}	ammonia feed rate per unit volume ($mol\,l^{-1}\,h^{-1}$)
H	Hamiltonian
L	leucine concentration ($g\,l^{-1}$)
p	product content of cells ($mg\,g^{-1}$)
P	product concentration (histidine or lysine) ($g\,l^{-1}$)
S	substrate concentration ($g\,l^{-1}$)
t	time (h)
t_c	switching time from μ_{max} to μ_c in the optimal profile (h)
V	working volume (l)
X	cell concentration ($g\,l^{-1}$)
Z	cell mass (g)
φ	slope of H with respect to μ
ψ	defined in Eq. (10)
λ_i	adjoint variable (i = 1, 2)
μ	specific growth rate (h^{-1})
ν	specific consumption rate (h^{-1})
ϱ	specific production rate (h^{-1})

Subscript

0	at initial time
C	critical or switching
E	ethanol
f	at final time
F	feed
G	glutathione
H	histidine
max	maximum
min	minimum
opt	optimal

1 Introduction

In a fed-batch culture, the substrate concentration can be maintained at a fairly low level and the unfavorable effects of a high concentration, such as growth inhibition, can be avoided. Because of this advantage, fed-batch cultivation techniques have been developed for many bioconversion processes. In such cases, depending on the cultivation process employed, the optimum feeding strategy of the substrate should be investigated so as to obtain the maximum production.

In this article, an approach for maximum bioproduct production in fed-batch cultures is explained, and its validity is demonstrated by experimental data. This is based on a model which describes the relationship between the specific production rate, ϱ, and specific growth rate, μ. Using a mathematical model, an optimal profile of the specific growth rate could then be easily obtained by the Maximum Principle. Finally, the optimal profile was realized by changing the feed rate of the substrate in a practical fed-batch culture.

As shown in several examples, the specific production rate, ϱ, had a maximum at a certain specific growth rate, μ_c, and the optimal profile was obtained as bang-bang control (boundary control). That is, in the early stage of the fed-batch culture μ should be kept at its maximum value, μ_{max} and for the next stage μ should be kept at μ_c. The control strategy can be comprehended as a two-stage production procedure composed of a cell growth stage and then a production stage. Practical examples of bioproduction, such as histidine and glutathione production, showed that the two-stage production process could be realized experimentally, thus demonstrating the validity of the method explained herein.

2 Optimization in Fed-Batch Bioreactors

2.1 An Approach for Maximum Production

To attain maximum production in a fed-batch culture, the three steps shown in Fig. 1 are usually passed through; i.e. modeling of the process, calculation of the optimal solution, and realization of the solution. Using this approach, any discrepancy between the model and the real process identified from the experiment

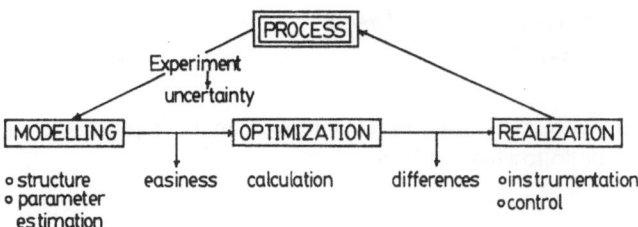

Fig. 1. Three steps for optimization of bioreactors

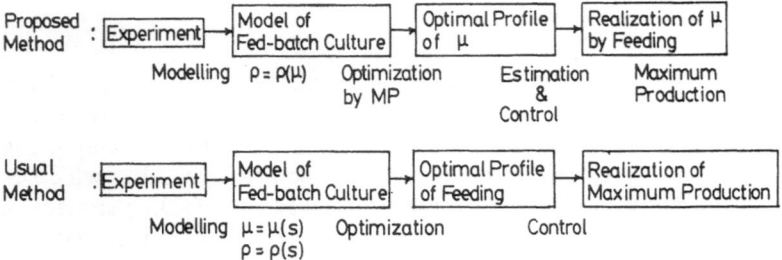

Fig. 2. Approaches for maximum production in fed-batch culture by the usual and proposed methods

is one of the problems taken into account. The ease of the optimization calculation is also another aspect which must be considered. From these viewpoints, one of the most appropriate approaches, in which an optimum solution is simple to obtain in spite of the fact that the above model can explain the data well, is proposed and represented in Fig. 2. In the modeling stage, one of the problems is how to quantitatively describe the reaction rates included in the mass balance equations. Usually, these rates have been described using the concentration of items such as the substrate, product and so on. Lim and his coworkers [1, 2, 3] have reported on optimization and control in fed-batch cultivation using this type of model, and they have successfully attained optimal operation. Based on the same type of model, Yamane' and Shimizu [4] proposed a calculation method of feed rate using the specific growth rate as an auxiliary parameter for ease of optimization.

However, here another approach is employed. First, the relationship between the specific growth rate and the specific production rate was identified and represented by a model. Then, the optimal strategy of the specific growth rate was obtained by the Maximum Principle [5]. Finally, this strategy was realized. Comparing this method with the usual approach to optimization, as shown in Fig. 2, a simple model with easy optimization but with a relatively complex realization method are the characteristics of the approach introduced here. Using the specific growth rate in this way, this method has been applied to several cultivation processes, as shown later, and it is expected to be applicable to other processes as well.

2.2 Modeling in Fed-batch Cultures Based on the Specific Growth Rate

A model of the reaction process in a fed-batch culture can be represented as follows. For cell growth, the following relation,

$$\frac{dVX}{dt} = \mu VX \tag{1}$$

can be derived, where V and X are the liquid volume and cell concentration, respectively. When the total cell mass, VX, is defined as Z, Eq. (1) can be rewritten as

$$\frac{dZ}{dt} = \mu Z, \tag{2}$$

where the initial condition of Z is given as $Z(0)$.

For accumulation of the bioproduct in the reactor, in the case of extracellular production

$$\frac{dVP}{dt} = \varrho(\mu) VX \tag{3a}$$

can be derived, where P is the product concentration $(g\,l^{-1})$. In the case of intracellular production

$$\frac{dVpX}{dt} = \varrho(\mu) VX \tag{3b}$$

is obtained, where p is the product content in the cell in $mg\,g^{-1}$.

By the feeding of substrate, the working volume of the reactor, V, increases in accordance with

$$\frac{dV}{dt} = F. \tag{4}$$

This modeling, based on the specific growth rate, is very simple but seems to be essential because all the biochemical reactions which occur are incorporated with cell growth. Of course, some exceptional cases, in which the specific production rate cannot be described directly using the specific growth rate, such as in secondary metabolite production, will also be met in practical cultivation processes.

2.3 Dynamical Optimization Problem

The maximum production problem in a fed-batch culture is to maximize the amount of product produced during a given operation time, t_f. The objective function, i.e. the amount of product accumulated, J, can be written by integration of Eq. (3a) or Eq. (3b) as

$$J = \int_0^{t_f} \varrho(\mu) Z\,dt, \tag{5}$$

where the total cell mas is written as Z instead of VX. Here, Z is subjected to Eq. (2).

Now, the specific growth rate, μ, is taken as the decision variable and its pattern is realized practically by changing the feed rate of the substrate. The optimization considered here can be summarized in cases of both intracellular and extracellular production as: to find the optimum profile of μ which maximizes J in Eq. (5) subject to Eqs. (2) and (4), with μ also subject to

$$\mu_{min} \leqq \mu \leqq \mu_{max} . \tag{6}$$

2.4 Theoretical Solution of the Optimization Problem

CASE (A): $\varrho(\mu)$ = a linear function of μ

In many cases, the relationship between μ and ϱ can be formulated approximately as two straight lines, e.g., as shown in Fig. 3, as

$$\varrho(\mu) = \begin{cases} a_1\mu + b_1 & \mu_{min} \leqq \mu \leqq \mu_c \\ -a_2\mu + b_2 & \mu_c < \mu \leqq \mu_{max} \end{cases}. \tag{7}$$

The mathematically stated problem can be solved theoretically by the Maximum Principle (5) as follows. The optimal profile of specific growth rate, μ_{opt}, can be obtained by maximization of the Hamiltonian, H, defined as

$$H \overset{\Delta}{=} \varrho Z + \lambda \mu Z \overset{\Delta}{=} \Phi \mu + \Psi , \tag{8}$$

where

$$\Phi = \begin{cases} (a_1 + \lambda) Z & \mu_{min} \leqq \mu \leqq \mu_c \\ (-a_2 + \lambda) Z & \mu_c < \mu \leqq \mu_{max} \end{cases} \tag{9}$$

$$\Psi = \begin{cases} b_1 Z & \mu_{min} \leqq \mu \leqq \mu_c \\ b_2 Z & \mu_c < \mu \leqq \mu_{max} \end{cases} \tag{10}$$

and λ is an adjoint variable represented by the following differential equation

$$\frac{d\lambda}{dt} = -\frac{\partial H}{\partial Z} = -\varrho - \lambda\mu \tag{11}$$

$$\lambda(t_f) = 0 . \tag{12}$$

Regarding Eq. (8), maximization of H results in boundary control which consists of a boundary value of μ such as μ_{min}, μ_c, and μ_{max} because H is a linear function of μ and the maximum value of H takes place at the boundary.

For finding the optimal pattern of μ, the following properties of Φ will be useful. For the gradient of Φ,

$$\frac{d\Phi}{dt} = -b_i Z < 0 \qquad i = 1 \text{ and } 2 . \tag{13}$$

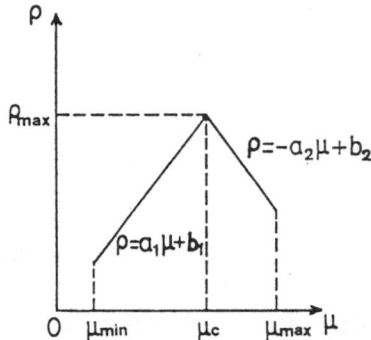

Fig. 3. Relationship between μ and ϱ.

At the final time, t_f,

$$\Phi(t_f) = \begin{cases} a_1 Z > 0 & \mu_{min} \leqq \mu \leqq \mu_c \\ -a_2 Z < 0 & \mu_c < \mu \leqq \mu_{max} \end{cases}.$$ (14)

Figure 4c shows the qualitative relationship between H and μ at $t = t_f$. Referring to Eq. (13), the relationship of H probably changes from that shown in Fig. 4a to 4b and finally to that of 4c. The optimal control, μ_{opt}, maximizes the Hamiltonian. Thus, it is given as the following boundary control (bang-bang control);

$$\mu_{opt} = \begin{cases} \mu_{max} & \text{for} \quad 0 \leqq t \leqq t_c \\ \mu_c & \text{for} \quad t_c < t \leqq t_f \end{cases}.$$ (15)

Here, t_c is the switching time of μ from μ_{max} to μ_c. At the swtiching time, t_c,

$$\Phi(t_c) = a_2.$$ (16)

By integration of Eq. (11), t_c can be obtained so as to satisfy the boundary conditions of Eqs. (16) and (12). The solution is

$$t_c = t_f - \ln\left(\frac{b_2}{b_2 - a_2\mu_c}\right)\Big/\mu_c$$

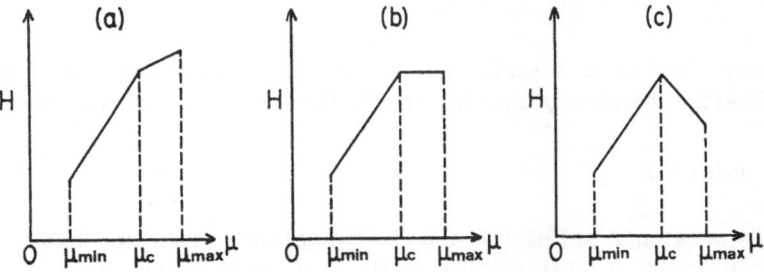

Fig. 4. Possible relationships between the Hamiltonian H and μ

or

$$= t_f - \ln \left(\frac{b_2}{\varrho_{max}} \right) \Big/ \mu_c , \tag{17}$$

where ϱ_{max} is the maximum value of ϱ. That is,

$$t_s \overset{\Delta}{=} t_f - t_c = \ln (b_2/\varrho_{max})/\mu_c . \tag{18}$$

It should be noted that the duration time, t_s, defined in Eq. (18) (for keeping $\mu = \mu_c$) was independent of the initial condition and t_f, and only dependent on the pattern of $\varrho(\mu)$.

In addition to the above discussion, if $t_f < t_s$, the duration of $\mu = \mu_c$ was optimum. Moreover, the singular extremum which will be derived from $\Phi \equiv 0$ is invalid because Eq. (13) should be satisfied. Then, it can be shown that the optimal solution never includes the singular extremum.

Finally, it should be noted that the strategy of the optimal pattern of μ suggested that a two-stage cultivation for growth and production was the best. The first stage was interpreted as cell growth and the second was for product formation.

CASE (B): $\varrho(\mu) =$ a nonlinear function of μ with an extremum

When the relationship between μ and ϱ cannot be approximated by a straight line, the optimal solution of the specific growth rate becomes similar to this case. Now let us consider the case where $\varrho(\mu)$ has an extremum at a certain μ, $\mu_c(\mu_{min} < \mu < \mu_{max})$. In this case, the local extreme condition; $\partial H/\partial\mu = 0$ can be satisfied. From $\partial H/\partial\mu = 0$,

$$\frac{\partial\varrho}{\partial\mu} = -\lambda(t) \tag{19}$$

is dervied. Then the optimal solution is given as

$$\mu_{opt} = \begin{cases} \mu_{max} & \text{for} \quad 0 \leq t \leq t_h \\ \mu_h & \text{for} \quad t_h < t \leq t_f \end{cases}, \tag{20}$$

where μ_h is obtained so as to satify Eq. (19). Of course, $\lambda(t)$ should satify Eqs. (11) and (12). At the final time, t_f, Eqs. (12) and (19) should be satified, and from this

$$\mu_h(t_f) = \mu_c \tag{21}$$

is derived. By integration of Eqs. (11) and (19) toward the reverse-time direction with the initial conditions of Eqs. (12) and (21), $\mu_h(t)$ can be calculated and also t_h can be obtained as the time when μ_h equals μ_{max}. For example, when ϱ is given as a second order polynomial such as

$$\varrho = \frac{\varrho_{max}(\mu - 2\mu_c)\,\mu}{2(\mu_c)^2}, \tag{22}$$

μ_h can be derived from Eq. (19) as

$$\mu_h = \frac{(\mu_c)^2}{\varrho_{max}}\lambda + \mu_c \quad \text{for} \quad t_h \leq t \leq t_f; \tag{23}$$

and λ is also obtained as

$$\left(t - t_f + \frac{2}{\mu_c}\right)\left(\frac{(\mu_c)^2}{\varrho_{max}}\lambda + \mu_c\right) = 2. \tag{24}$$

Just at $t = t_h$, μ_h should be equal to μ_{max} and from this

$$t_f - t_h = 2\left(\frac{1}{\mu_c} - \frac{1}{\mu_{max}}\right). \tag{25}$$

In this case, $(t_f - t_h)$ is independent of the initial conditions and t_f in the same manner as *CASE (A)*.

In addition to these, optimal solutions for other cases: 1) where $\varrho(\mu)$ is a monotonically increasing function, and 2) where $\varrho(\mu)$ is a monotonically decreasing function, can be also obtained. For a monotonically increasing function of μ, the solution is trivial, as $\mu = \mu_{max}$ for all the operation time. For case 2), the solution is similar to *CASE (B)* except that $\mu = \mu_{min}$ will continue depending on the conditions of the problem ($\mu = \mu_c$ is maintained for only a moment in *CASE (B)*).

CASE (C): Other types of optimization problems

In some cases, two objectives should be evaluated as follows.
(1) The amount of total product, that is the product of cell mass, Z, and product content, p, should be maximized.
(2) The product content, p, should be kept at a high value at the final time of the cultivation because of economic factors in downstream processing.

Considering the above two objectives, the optimization problem can be stated thus: to find the optimal profile of the specific growth rate, μ, so as to maximize the total amount of product and to keep the product content at a high value at the fixed final time.

The optimization problem can be formulated mathematically as follows: The problem is to find the optimal profile of μ so as to maximize the following performance index,

$$\begin{aligned}
J &= p(t_f)\,Z(t_f) \\
&= p(0)\,Z(0) + \int_0^{t_f} \varrho_G Z\,\mathrm{d}t\,, \tag{26}
\end{aligned}$$

under the constraint of the final time as

$$p(t_f) = p_f \tag{27}$$

in a fed-batch culture described by Eqs. (2), (3b) and (4), where t_f and p_f are the fixed final time of the cultivation and the product content in the cell at the final time, respectively. Further, μ was subject to Eq. (6) and initial conditions were imposed as follows:

$$Z(0) = Z_0 \tag{28}$$

$$p(0) = p_0, \tag{29}$$

where Z_0 and p_0 are initial values of cell mass and product content in the cell, respectively.

According to the maximum principle, the optimal profile of the specific growth rate, μ_{opt}, can be obtained by maximization of the Hamiltonian, given as

$$H = \varrho_G(\mu)\, Z + \lambda_1 \mu Z + \lambda_2 \{\varrho_G(\mu) - \mu p\}, \tag{30}$$

where λ_1, λ_2 are adjoint variables represented by the differential equations

$$\frac{d\lambda_1}{dt} = -\varrho_G(\mu) - \lambda_1 \mu \tag{31}$$

$$\frac{d\lambda_2}{dt} = \lambda_2 \mu \tag{32}$$

with the terminal conditions

$$\lambda_1(t_f) = 0 \tag{33}$$

$$\lambda_2(t_f) : \text{free} . \tag{34}$$

According to the considerations for every possible situation, it is again concluded that the optimal profile of the specific growth rate is a bang-bang one, which is described by one of the three following alternatives:

1) $\mu_{opt} = \mu_{max}$ (for $0 \leqq t \leqq t_f$)   (35)

or

2) $\mu_{opt} = \mu_{max}$ (for $0 \leqq t \leqq t_B$)

$\mu_{opt} = \mu_c$ (for $t_B < t \leqq t_f$)   (36)

or

3) $\mu_{\mathrm{opt}} = \mu_{\mathrm{c}}$ (for $0 \leqq t \leqq t_{\mathrm{f}}$).   (37)

For the more general cases as given in Eq. (36), μ should at the maximum value, μ_{max}, and should be kept at μ_{max}. Then, μ should be switched to μ_{c}, which gives a maximum value of ϱ_{G}. The switching time, t_{B}, depends on the final required value of the product content, $p(t_{\mathrm{f}})$.

3 Control in Fed-Batch Bioreactors

3.1 Role of the Control of the Specific Growth Rate

Apart from the approach proposed here, the specific growth rate, μ, is one of the important process parameters which represent the dynamic behavior of the bioreactor. For example, in a baker's yeast bioreactor, it is desired that μ should be maximum during cultivation in order to obtain maximum cell production. For maximum cell production, the sugar concentration in the medium should be kept at an optimal value. However, there is no industrially available glucose sensor for use on-line. Thus, instead of measuring and controlling the glucose concentration, RQ and the ethanol concentration are usually controlled (6, 7). However it should be stressed that a control system that functions by controlling RQ and the ethanol concentration can be utilized only for maximization of the specific growth rate. When an arbitrary value of μ or some particular pattern of μ, e.g., the bang-bang type profile control of μ which is required for the quality control of baker's yeast (8), is desired, another control scheme will be necessary.

To maintain μ at a constant value in a fed-batch culture, an exponential feeding policy is frequently used. However, if the initial condition and parameters which are needed for calculating the feed rate contain errors, the specific growth rate is not equal to the desired value at all. That is, in many cases a closed-loop system is needed for controlling the specific growth rate.

There are few reports which show that the specific growth rate can be controlled at any set-point value or prescribed profile by a feedback controller. In our previous simulation study (9), a proposed Programmed controller/Feedback compensator (PF) system was successfully applied to the profile control of μ where it was assumed that μ could be measured directly. Also, our experimental study (10) showed a) the availability of the PF system for the profile control of μ, which is estimated online by the extended Kalman filter, and b) the usefulness of the PF system for achieving quality control of the baker's yeast. In this section, the PF system is briefly described and the problem of estimating the specific growth rate (11), which is usually inevitable for this system, is described.

3.2 The PF System

To obtain a sophisticated control system, first, the PID feedback control system was analysed so as to clarify the role of the individual parts of the PID control action, and a type of programmed/feedback controller was developed (12). As a result of the findings, a new control scheme for fed-batch culture was designed (9, 10, 13). This scheme, called a "Programmed controller/Feedback compensator (PF) System", consists of a programmed controller and a pre-compensator as shown in Fig. 5. The specific growth rate, μ, should follow the desired profile under the control of the programmed controller unless there is noise or disturbance, which should be compensated for by the pre-compensator. As a pre-compensator, the Model Reference Adaptive Control (MRAC) algorithm was proposed, because a cultivation process is a time variable and highly nonlinear system, and the whole system was named PF-MRAC. Numerous computer simulations have verified the usefulness of the control system (9). However, in many cultivation processes as shown later, simple PI control can be effectively used as a pre-compensator.

The nominal feed rate of the substrate, $\bar{F}$, is determined by the programmed controller as follows. The dynamics of the cell concentration, X, and the substrate concentration, S, can be generally formulated as

$$\frac{dX}{dt} = \left(\mu - \frac{F}{V}\right) X \tag{38}$$

$$\frac{dS}{dt} = -\frac{\mu X}{Y} + \frac{F}{V}(S_F - S), \tag{39}$$

where Y and S_F are the yield coefficient of cell growth and the substrate concentration of the feed, respectively. If μ can be kept at a constant value, μ^*, for a certain period, the substrate concentration, S, must be constant based on the assumption that μ is a function of S. And if the desired profile of μ^* consists of such piecewise constant values, the nominal feed rate of the substrate, $\bar{F}$, can be given as

$$\bar{F} = \frac{\mu^* X V}{Y(S_F - S)} \tag{40}$$

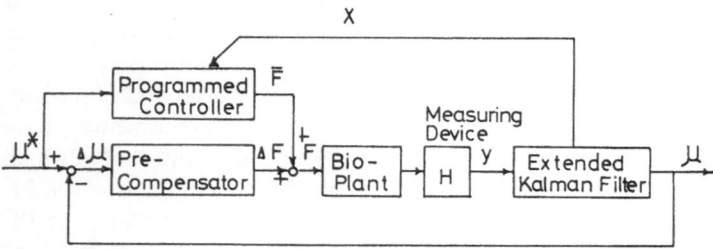

Fig. 5. PF system for controlling μ

because the left hand side of Eq. (2) should be equal to zero, where μ^* is the objective value of the specific growth rate. If it can be assumed $S_F \gg S$, Eq. (40) becomes

$$\bar{F} = \frac{\mu^* X V}{Y S_F} . \tag{41}$$

In order to calculate $\bar{F}$ using Eq. (41), X and V should be given each time. However, if the given initial value is different from the real one and μ is not controlled as well as desired, the calculated value of X may have a large error. In order to improve on this disadvantageous point, X was estimated by the extended Kalman filter; this estimated value of X will be the most probable on. Furthermore, in Eq. (41), V, for the fed-batch culture is given as

$$V = V_0 + \int_0^{t_f} F(\tau) \, d\tau . \tag{42}$$

According to this programmed controller, the nominal feed rate, F, can be given adaptively depending on the cell concentration. Of course, to compensate for unexpected disturbances or uncertainties, the pre-compensator should work successfully.

3.3 Estimation of the Specific Growth Rate

In order to construct a feedback profile control system for the specific growth rate, the specific growth rate itself must be observed or estimated. Furthermore, the value of the cell concentration should be given accurately so as to correct the nominal feed rate of the substrate each time. To estimate these values, the extended Kalman filter was utilized, a technique (14) that has sometimes been used for on-line state estimation in cultivation processes (15, 16, 17, 18).

It has been proposed that the overall growth rate can be estimated indirectly by elemental and macroscopic balances of the reactor utilizing other measurable rates (19, 20). Because the final objective is control of the specific growth rate, it is required that the estimated value should converge with the true one as quickly as possible, though the linear Kalman filter theory guarantees convergence only in the infinite stage. Also, if the estimation error is large, the feedback controller cannot compensate for disturbances but instead produces new disturbances. Therefore, the convergence of the specific growth rate with the objective value by the feedback controller is strongly affected by the accuracy of the estimation.

The indirect measuring and on-line estimation methods in bioreactors pose many problems to be considered in their application, namely:

1) What variable should be observed to avoid singularity of the algebraic linear equation in the macroscopic material balancing method?
2) What type of system equation should be given to the extended Kalman filter?
3) How should the covariance matrix of the system noise be given?

4) How should the initial values of the state variables which cannot be measured directly be given?

5) How can the accuracy of the on-line estimated value be evaluated based on the observed data?

These problems will be encountered frequently in any state estimation related to bioprocesses. However, the answers cannot readily be given in absolute terms. It will be easily understood that optimally tuned parameters strongly depend on the kind of micro-organisms and sensors of the cultivation system; an algorithmic approach will therefore be required so as to get weel-tuned parameters. There are a number of studies concerning on-line state estimation, but there are few reports in which the on-line estimation procedure is applied to real processes (15, 21). Even in these reports, there are no algorithms to give information like that noted above, such as a measurable evaluation method for the covariance matrix of noise.

The sensitivity of parameter identification to small changes in the respiratory quotient is a big problem, as Grosz et al. have already shown (22). They studied singularity in macroscopic balances used in connection with respiratory quotient (RQ) measurement and heat evolution measurement in yeast and *E. coli* cultivation, and proved the existence of singularity when RQ is close to 1 in yeast cultivation with glucose and ammonia as substrate, confirming this by simulations and experiments. In our own report (11), a convenient and efficient numerical method which can easily find the singularity with no biochemical assumptions or information, was described. The method can originally measure the singularity of the coefficient matrix of a linear algebraic equation.

In applying the Kalman filter to the estimation of the specific growth rate, the determination of a required system equation describing the dynamics of the specific growth rate is very difficult. Usually the specific growth rate is represented by a function of the substrate concentration, such as Monod's model. The parameters of such models are frequently identified from the data of the steady state chemostat culture. However, the identified parameters are frequently changed from one batch to another batch or in fed-batch culture. Stephanopoulos and San made a dynamical model of the specific growth rate as colored noise (15), but since their model has no biological background it cannot be judged whether it can represent exactly the dynamics of the specific growth rate in any cultivation system. The appropriate system equation of the extended Kalman filter should be sought in each process. An algorithm to search for the system equation of the specific growth rate will thus be necessary.

When the parameters required for the extended Kalman filter, such as the covariance matrix of the system noise and the initial values of the state variables, are going to be tuned, the criteria of the accuracy of the estimation should be given rationally. Criteria for the estimation of the specific growth rate have been proposed (11). Based on these, a system equation for the specific growth rate was selected and the initial values of the state variables and covariance matrix of the system noise were adjusted. From many experiments, it was confirmed that the specific growth rate in batch or fed-batch cultivation could be estimated accurately by means of the algorithm proposed. The adaptive filter (14) and the extended Kalman filter using the covariance matrix with a constant element were compared and evaluated based on the proposed criteria (11).

System Equation. When the overall growth rate, μX, or the cell concentration, X, is measured, μ should be estimated based on the online data by the extended Kalman filter. For this, a system equation which describes the dynamic behavior of μ should be introduced. With less information about the system, it is desirable that μ should be estimated in any case. From this viewpoint, the following equations for μ,

$$\frac{dX}{dt} = \left(\mu - \frac{F}{V} \right) X + \eta_1 \tag{43}$$

$$\frac{d\mu}{dt} = C + \eta_2 \tag{44}$$

$$\frac{dC}{dt} = -\frac{F}{V} C + \eta_2 \tag{45}$$

have been proposed by Stephanopoulos and San (15, 23), where C is a new variable introduced here, and η_i is white noise.

Using these equations and the output data of y which can be written by

$$y = \mu X + \xi_1 \tag{46}$$

or

$$y = X + \xi_2, \tag{47}$$

the extended Kalman filter can be applied for the estimation of the specific growth rate, μ, where η_i and ξ_i (i = 1, 2) are white noises.

Estimation by an Observer. Another estimation scheme based on an observer will be utilized for estimation of μ. An adaptive nonlinear observer for the estimation of cell concentration and the specific growth rate has been proposed by Dochain and Bastin (24). For their observer, the estimated value of cell concentration $X(t)$ and the estimated value of the specific growth rate $\mu(t)$ are given by the following ordinary differential equations;

$$\frac{dX(t)}{dt} = \left\{ \mu(t) - \frac{F(t)}{V(t)} + \lambda_1 (Xm(t) - X(t)) \right\} Xm(t) \tag{48}$$

$$\frac{d\mu(t)}{dt} = \lambda_2 Xm(t) (Xm(t) - X(t)) \tag{49}$$

$$\frac{dV}{dt} = F, \tag{50}$$

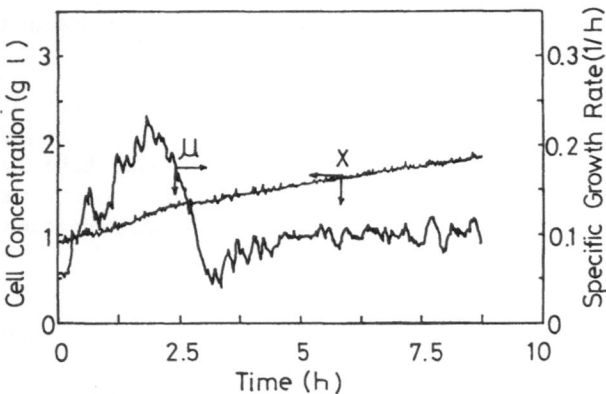

Fig. 6. Estimated value of μ by the extended Kalman filter (8)

where $Xm(t)$ is the measured value of X at time t, and λ_1 and λ_2 are positive parameters to be tuned from experiments with the cultivation process. The correction terms $\lambda_1(Xm(t) - X(t)) Xm(t)$, $\lambda_2 Xm(t) (Xm(t) - X(t))$, in above equations are adjusted by parameters λ_1 and λ_2. Their new idea is on that the correction terms are taken as the product of the estimation error $(Xm(t) - X(t))$ and $Xm(t)$.

Example of the Estimation of μ. As shown in Fig. 6, μ is estimated by an extended Kalman filter based on the on-line data of the cell concentration, X. X is also estimated by filtering noise. Many experiments have verified that the extended Kalman filter proposed previously could be applied to the estimation of the specific growth rate and cell concentration successfully in baker's yeast fed-batch culture (11, 13). Also, the nonlinear observer described here could be used for estimation successfully (11).

4 Application of the Approach to Batch and Fed-Batch Bioreactors

4.1 Maximum Histidine Production in a Fed-batch Culture

Histidine is one of the amino acids industrially produced by bioprocesses. The improvement of the cultivation conditions is one of the important measures used to increase productivity. *Brevibacterium flavum* FERM 1564 (Japan Kokai Tokkyo Koho, JP 52-21595, Cl.c12-D13/06, 5 p., 1977) was used as a histidine-producing mutant. The characteristics of this microorganism have already been analyzed as follows (25): (a) The microorganism was a uracil auxotrophic mutant and $250\ mg\ l^{-1}$ of uracil in the medium was optimum. (b) A relatively high DO concentration was required to avoid the production of undesirable amino acids. Consequently, the DO concentration was controlled at about 6 to 7 $mg\ l^{-1}$ using

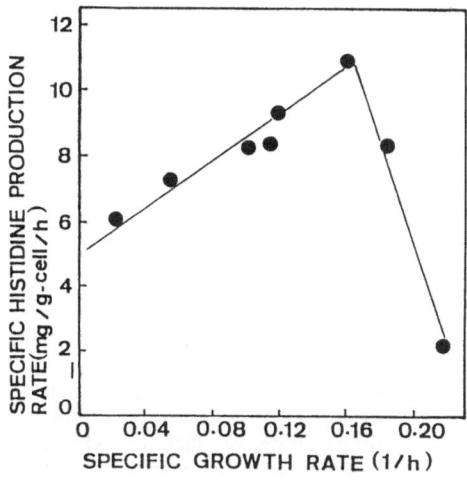

Fig. 7. Relationship between specific histidine production rate and specific growth rate (23)

pure oxygen. (c) The optimum A/G consumption ratio for the maximum specific histidine production rate in the fed-batch culture was decided to be around 2.4.

Figure 7 shows the relationship between the specific growth rate, μ, and the specific histidine production rate, ϱ_H (26). When μ was less than 0.16, the ϱ_H increased proportionally to μ but decreased suddenly when μ was larger than 0.16. That is, the ϱ_H had a maximum at a certain value of $\mu = \mu_c$ (0.16 h^{-1}). The decrease of ϱ_H might be caused by an imbalance of energy supply for growth and production. The parameters a_i and b_i in Eq. (7) were estimated as 43.14, 147.71 and 4.2, 34.72, respectively by using the least square method. In this case, $t_s = 7.3$ h was suggested as the optimal duration time of $\mu = \mu_c$ from the experimental data of Fig. 7.

Figure 8 shows one of the results of the optimal feeding strategy in the case of $t_f = 10$ h. The control or the realization method employed here was relatively simple. That is, exponential feeding (feedforward) control was utilized but the initial condition, $V_o X_0$, was renewed whenever the cell concentration (determined every hour) did not follow the expected value. A broken line corresponds to the desired trajectory of μ and a solid line is the actual μ calculated from the off-line data of $Z(= VX)$ using the extended Kalman filter. The feed flow rate of the substrate is also illustrated. The actual μ did not always coincide with the ideal optimal trajectory because of a relatively poor controller performance. An improvement in the controller's performance for μ should be aimed at, considering the long time delay of the system dynamics. The objective function, J, is also shown in this figure. The data for non-optimal control of feeding in which the desired μ was taken as μ_c (0.16 g^{-1}) or μ_{max} (0.2 h^{-1}) are illustrated for comparison with the optimal control data. Since the feeding times of these two data were not fixed at 10 h, but were 9 and 8 h, respectively, their trends toward 10 h are indicated by extrapolation. To compensate for the difference of initial cell mass between the three experiments, every piece of data for J was recalculated by multiplication proportionally to the difference of initial cell mass based on the optimal case. The

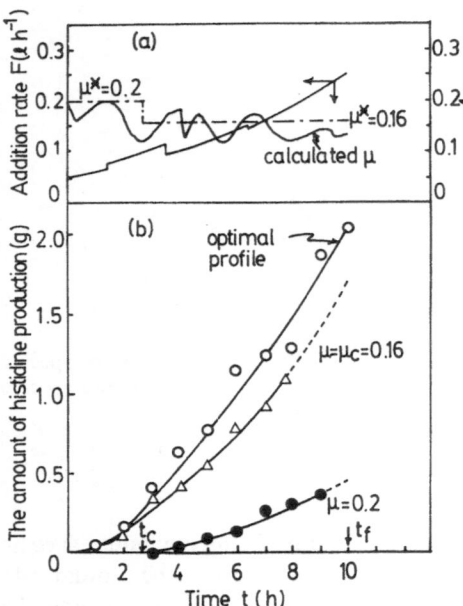

Fig. 8. The result of maximum histidine production in fed-batch culture; total histidine production ($\circ$) by bang-bang control for $\mu = \mu_c = 0.16 \text{ h}^{-1}$ and ($\bullet$) by control for $\mu = \mu^*$, ($\triangle$) by constant control for $\mu = \mu_{max} = 0.2 \text{ h}^{-1}$ (23)

result indicated that the difference between μ_c and μ_{opt} was not so significant, though J_{opt} by the operation of μ_{opt} was still larger than that obtained in other cases. Thus, Fig. 8 exhibits the excellence of this method and suggests that a two-stage production process of histidine is useful in fed-batch culture. However, it should be stated again that the realization step for maximum production will be possibly improved if a more sophisticated control strategy of μ is taken into account.

As already explained, the usual approaches for maximization of histidine production were extremely difficult to accomplish because of two big problems encountered in model building. One was due to the double substrate (glucose and acetate) limitation for growth and histidine production, while the other was caused by the difficulty in measuring and developing the model of specific growth rate and specific production rate in the range of low concentrations of glucose and acetate. This is because an accurate measurement of these concentrations is difficult to attain. Nevertheless, using the approach proposed here, the maximum production of histidine in fed-batch culture was partially accomplished; this approach could be applicable to the production of other bioproducts by fed-batch culture.

4.2 Enzyme Production in a Yeast Batch Culture with Temperature-Controlled Gene Expression

In the approach to optimization in fed-batch or batch culture based on the specific growth rate explained here, the modeling and optimization are relatively easy, but realization of the optimal profile obtained in the previous step is rather difficult, and as has been shown, sometimes requires advanced techniques. The problem of controlling the specific growth rate so as to follow the specified optimal profile of μ, usually by manipulating the substrate feed rate, sometimes encounters difficulties as already explained. This will also be referred to again later.

Here, as an example of the rather easy control of the specific growth rate by cultivation temperature shift, enzyme production by yeast cultivation with temperature-controlled gene expression, will be discussed. Let us consider the maximum production of acid phosphatase (abbreviated as APase) in *Saccharomyces cerevisiae* using the host-vector system, in which the cultivation temperature can control the expression of the transferred gene. The 2μ-DNA plasmid harboring the MFα1 promotor with the APase gene was transfected into the temperature-sensitive host, *Saccharomyces cerevisiae*, which expresses the α-type at over 30 °C. That is, at over 30 °C, the MFα1 promotor expresses, and simultaneously the APase gene produces APase. In contrast, at temperatures lower than 30 °C, APase is not produced. It is not only the amount of APase gene expression, but also the specific growth rate, which depends on the cultivation temperature.

The experimental results showing the relation between the specific growth rate or specific production rate and cultivation temperature are summarized in Fig. 9. The problem of maximum production of APase during batch cultivation can also be comprehended as how to compromise between growth and production. The specific rates of both items, growth and production, depend on the cultivation temperature and total amount of product; the objective function employed here will be directly affected by growth and production in the same situation as already explained.

Rearranging Fig. 9 into Fig. 10 and formulating an optimization, the solution can be easily obtained also as bang-bang control of the specific growth rate. That is, μ should be changed from μ_{max} to μ_c at a certain switching time. Again, the duration for which μ_c is kept can be fixed independent of the operation time, t_f. Here, μ

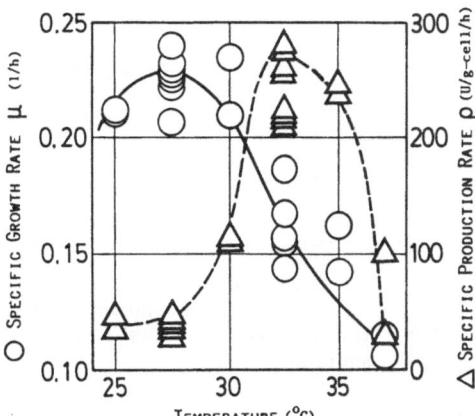

Fig. 9. Effect of cultivation temperature on specific APase production rate and specific growth rate

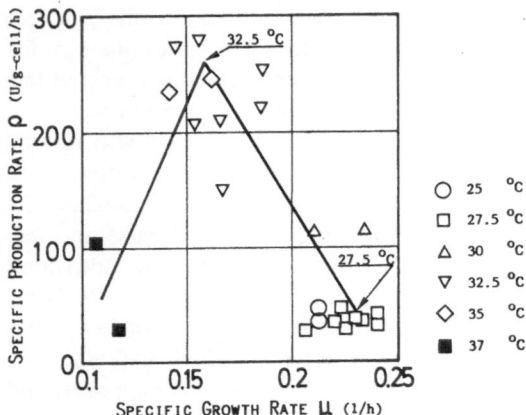

Fig. 10. Relationship between specific APase production rate and specific growth rate

can be controlled well by changing the cultivation temperature. As expected, the control of μ by temperature shift is easier than by substrate feed rate change.

Fig. 11 shows the relationship between the switching time and product APase concentration. In batch cultivation, the amount of the total product completely corresponds to the product concentration because of the constant working volume. The open triangles show the experimental data of the final APase concentration and the solid line represents the calculated APase concentration taking into account the fact that the expression of APase gene stopped 5 h after the temperature was shifted up. The broken line shows the concentration when this fact was not considered; the arrow indicates the optimal switching time, which can be calculated from Eq. (17).

As can be seen in Fig. 11, the calculated and experimental data coincided well. This means that the temperature shift (the cultivation temperature can be easily shifted up if the working volume is not so large) can control the specific growth rate well. Thus, a sophisticated estimation and control system for the specific growth rate will not be needed, at least in terms of a procedure to realize the optimal profile for this cultivation.

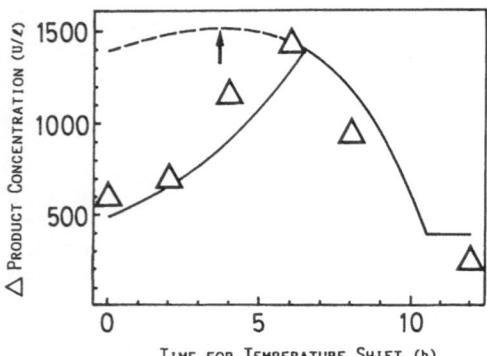

Fig. 11. Relationship between switching time (time for temperature shift) from μ_{max} to μ_c and final APase concentration

4.3 Maximum GSH Production in a Yeast Fed-Batch Culture

Glutathione, abbreviated as GSH, is a medically active tripeptide consisting of glutamate, cysteine and glycine, which widely affects oxidation and reduction in vivo. Recently, GSH has been reported as a medicine for the liver and as a scavenger of toxic compounds. It is well known that the GSH content in a certain strain of the yeast is high. It can assimilate glucose and also produce ethanol due to the Crabtree effect when the glucose concentration in the bioreactor is very high. Also, it has been reported that the productivity of GSH depends upon the kinds of carbon sources.

The relationship between μ and the specific glutathione production rate ϱ_G can be summarized as shown in Fig. 12. The relationship between μ and the specific production rate of ethanol ϱ_B is also shown. It should be noted that ϱ_G had the maximum value when μ was $0.30\ h^{-1}$. Ethanol was produced at higher μ values than $0.30\ h^{-1}$ by the Crabtree effect. This critical value of μ coincided completely with the value at which ϱ_G had the maximum; ϱ_G decreased when ethanol was produced, due to high glucose concentration (27).

The optimal profile of μ for the maximum production of GSH was actually realized by controlling the feed rate of the substrate. At first, μ was indirectly controlled by exponential feeding plus ethanol concentration control. One of the control results formulated by Eq. (36) is shown in Fig. 13. The optimal switching time, t_B, for the problem *CASE (C)*, as already explained, is calculated as 3.5 h for $p(t_f) = 11$ mg GSH per of cells and $t_f = 10$ h. The dotted line in the figure represents the desired trajectory of μ and the solid line is the real value of μ estimated by the extended Kalman filter. Exponential feeding of the substrate was undertaken for the first 2.5 h. Because of the delay of the response of μ against

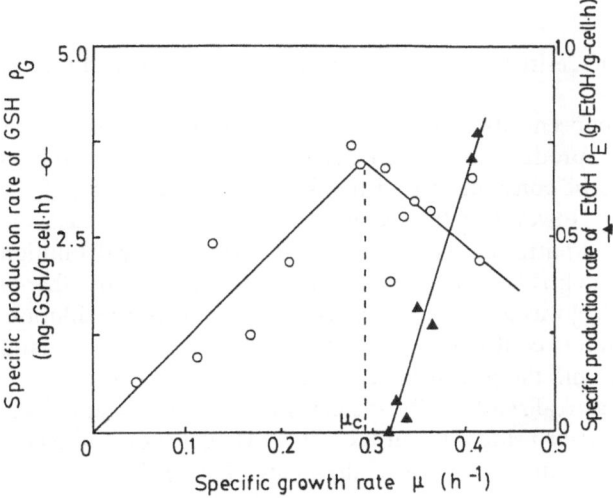

Fig. 12. Relationship between μ and ϱ_G ($-\circ-$), ϱ_B ($-\blacktriangle-$) (24)

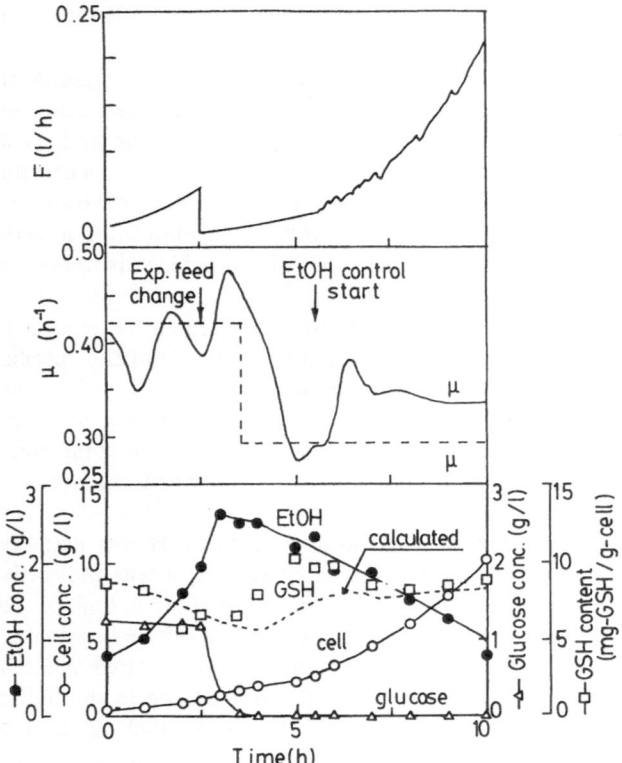

Fig. 13. Time course of state variables in the bioreactor when μ was controlled to the optimal profile using exponential feeding plus ethanol concentration control (24)

the change of feed rate, the desired μ for the exponential feeding was changed to the new one at 2.5 h.

After 5 h, the ethanol concentration was controlled by feeding the substrate. The yeast should neither produce nor assimilate ethanol at μ_c. The ethanol concentration should thus be constant at μ_c if the dilution by the feeding of the fresh medium is negligible. However, in this experiment this was not the case. On the contrary, the ethanol concentration should decrease in accordance with dilution due to the feeding if μ is controlled at μ_c. Thus, the objective value of ethanol concentration for the control was given as a decreasing curve due to the dilution. If the ethanol concentration is controlled to the desired decreasing curve, μ and ϱ_G can be controlled at μ_c and the possible maximum value, respectively.

This control strategy was realized by a PF system. Ethanol concentration was measured by an ethanol sensor using a teflon tube and FID detector. Indirectly, μ was controlled at μ_c successfully, as seen in the figure. The GSH content, p, increased after switching μ from μ_{max} to μ_c, as expected from the model. In the figure, the broken line shows the calculated values of the GSH content using the

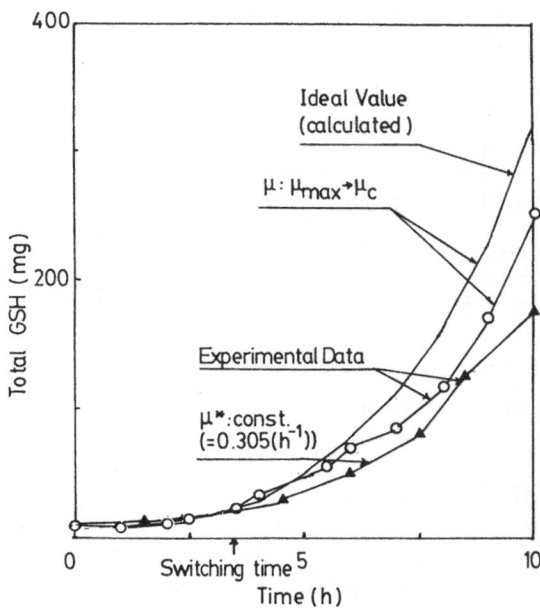

Fig. 14. Comparison of total GSH by optimal control of μ with that by constant μ (24)

mathematical model already explained in Eqs. (1), (3b) and (4) where $\varrho(\mu)$ is evaluated from Fig. 13. The calculated values almost coincide with the measured values except for the period when μ had a large stepwise change.

A comparison of the time courses of the amount of total GSH between constant μ and bang-bang control of μ was performed, as shown in Fig. 14. Open circles indicate the time course of the total amount of GSH obtained by the optimal profile control of μ. On the other hand, the closed triangles indicate the total amount of GSH obtained by constant control of μ. The constant value of μ was 0.305 h^{-1}, which is very close to μ_c. As shown in the figure, the total GSH obtained by the optimal strategy was about 1.41 times larger than that obtained by constant control of μ. The calculated expected values of total GSH are also shown.

The maximum production of GSH was also accomplished by controlling μ to the optimal profile directly, with the extended Kalman filter and PF System. Here, the optimal switching time is calculated as $t_c = 3.95$ h for $t_f = 6$ h from the solution of the *CASE (A)* problem. The estimation and control system of μ is shown in Fig. 15. Here, μ could be estimated as follows. The nitrogen uptake rate by the cell, R_{NH3}, could be calculated from the ammonia feed rate, F_{NH3}, for pH control considering the material balance of the nitrogen. The growth rate, μX, could be estimated from the relationship between F_{NH3} and μX. Moreover, μ and the cell concentration, X, were estimated by the extended Kalman filter, using μX as the measured variable. Estimated μ could be controlled by the PF System. The standard nominal feed rate, $\bar{F}$, calculated as exponential feeding, was given by the programmed controller, and the actual feed rate, F, was given as the sum of $\bar{F}$ and ΔF calculated by the pre-compensator.

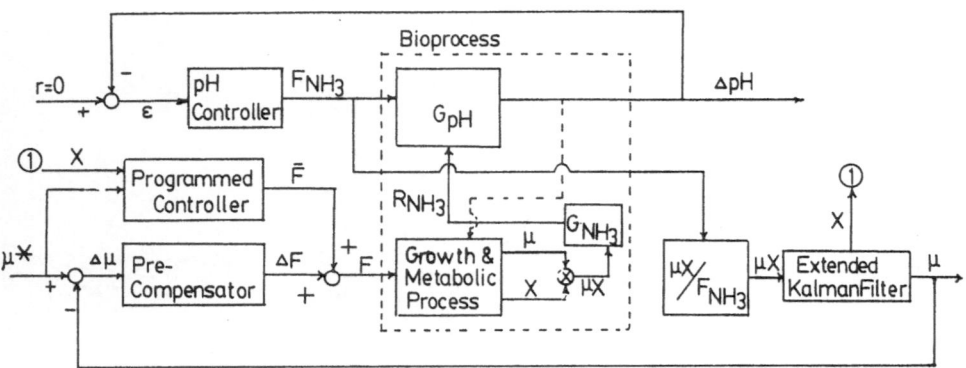

Fig. 15. Structure of the estimation and control system of μ

The estimation and control results for the maximum production of GSH under
6 h operation is shown in Fig. 16. The dotted line in the feed rate figure stands
for the standard feed rate, $\bar{F}$, while the solid line stands for the actual feed rate,
F. The dotted line in the μ figure is the desired profile of μ, while the solid line
is the actual one. For the maximum production of GSH without any restriction

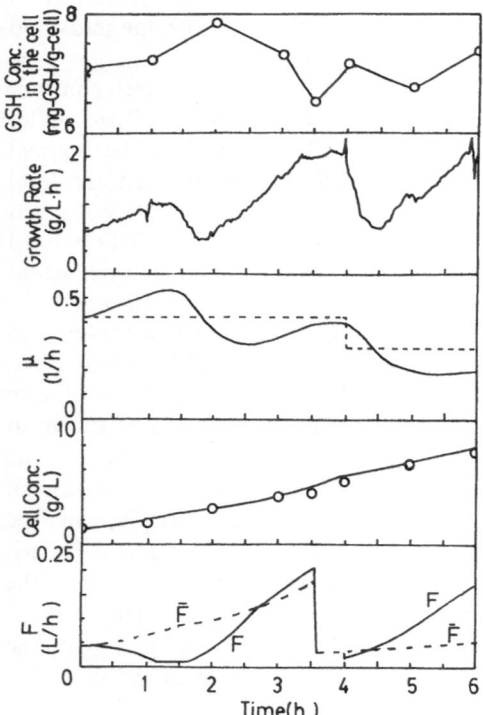

Fig. 16. Results of estimation and
optimal profile control of μ

for $p(t_f)$, the switching time t_c was given as a time 2.05 h before the final time t_f, as mentioned before. The solid line indicates the estimated values by the extended Kalman filter. Open circles and the solid line in the cell concentration figure stand for off-line data and estimated values by the extended Kalman filter, respectively. Estimated value of cell concentration accurately coincided with the measured ones and it was confirmed that μ and the cell concentration could be estimated by the extended Kalman filter satisfactorily; μ was controlled according to the optimal trajectory by the PF System.

A comparison of the total amount of GSH for both control strategies was performed, as shown in Fig. 17. Open circles are the time course of the total GSH, pZ, obtained when μ was controlled to the optimal profile. On the other hand, the closed triangles indicate the time course of the total GSH by constant control of μ. The total amount of GSH at the final time by optimal control of μ is more than that of the constant control of μ.

As shown in Figs. 13 and 16, the specific growth rate, μ, could not be well controlled. Also, the total amount of GSH produced was much less than the expected value calculated using the optimal profile of μ, as shown in Figs. 14 and 17. The reasons for these differences are mainly: 1) The relationship between μ and ϱ_G shown in Fig. 12 was obtained from data where the specific growth rate was constant (quasi-steady state) in the exponential fed-batch culture, and it was assumed that the relationship was valid when μ was not constant. However, this

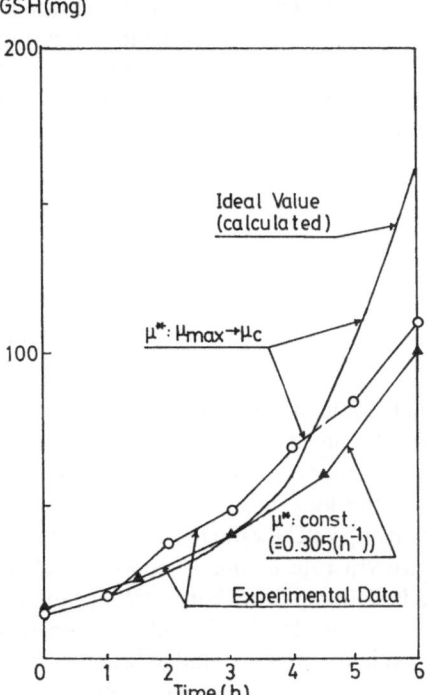

Fig. 17. Comparison of total GSH obtained by optimal control with that obtained by constant μ

assumption would not be acceptable in the practical control experiment. 2) The optimal profile of μ could not be realized perfectly — only roughly. The calculated values of GSH content using the relationship in Fig. 12 are shown as dotted lines in Fig. 13. The calculated content of GSH was not constant, even if μ was constant (that is, ϱ_G was constant), because of the difference in the initial conditions of GSH content. Anyway, the assumption mentioned above may not always be satisfied. In our optimization problem, the suggested optimal profile of μ is bang-bang control, which is composed of constant control of μ. Then, if control of μ is perfect, the first problem can be eliminated. It is, after all, improvement of the controller performance for μ which is most important in eliminating the gap between the theoretical and practical amounts of GSH produced.

For controlling the specific growth rate, the dynamic response according to the change of F plays an essential role, and the controller performance depends on the system dynamics and the employed controller itself. In other words, the extent of the improvement by the controller is limited by the system dynamics.

No mathematical model was used for describing these dynamics. (This is one of the prominent characteristics of our approach.) Thus, it is rather hard to evaluate whether the controller performance is acceptable or not considering the system dynamics. From our experience, the controller performance in Fig. 13 is fairly good, and in Fig. 16 it is good. Probably, the estimation procedure of μ affects the controller performance. This point will be improved by further investigation into the coordination between the estimator and the controller.

The total amount of GSH produced was, nevertheless, still larger than that obtained by the constant control of μ, as shown in Figs. 14 and 17, even enough the controller performance is not necessarily satisfactory. Further investigation will be focused on coordination between the estimation and the control of μ, and an improvement of the controller performance or analysis of the limits of the controller. This will make the approach proposed here a more interesting and valuable one for other cultivation processes.

5 Evaluation of the Approach

5.1 Modeling Error and Optimality

An approach for the optimization and realization of maximum production in fed-batch or batch cultures has been presented. One claim for this approach is in the modeling. The relationship between the specific growth rate and specific production rate is frequently obtained from the long term data of exponential fed-batch cultures. This means that the relationship will be valid for the quasi-steady state. During an experiment, the specific growth rate is almost constant. Thus, when the relationship between μ and ϱ includes a dynamic response, the optimality for the solution obtained from this approach will also be doubtful. From this point of view, the following example is given in order to judge whether this approach is acceptable from the practical viewpoint. The approach is very simple, and the discrepancy from the optimal strategy is not large but acceptably small for practical operation, even if the model is not rigorously correct but an approximated one.

5.2 Maximum Lysine Production by a Leucine-Limited Fed-Batch Culture

An auxotrophic mutant is usually used for the industrial production of amino acids for high productivity. However, the auxotrophic substrate which is required for the growth of the mutant frequently inhibits or represses the production of amino acid. Hence, there exists an optimal concentration for the maximum production of amino acid in a fed-batch culture.

In this section, lysine production by *Corynebacterium glutamicum* which is an auxotroph and requires leucine for cell growth, is discussed as an example of a cultivation process. From several experiments, it was shown that leucine repressed the specific production rate of lysine. However, the optimal leucine concentration could not be found because the usual type of analyzer, such as an amino acid auto-analyzer, could not detect the very low concentration of leucine which must be optimal. In order to obtain the optimal leucine concentration, we carried out many exponential fed-batch cultures which were regulated by leucine limitation. By controlling the specific growth rate in the exponential fed-batch culture, the relationship between μ and ϱ was identified.

At first, we built a simple model (Model-1) which describes the relationship between μ and ϱ, as shown in Fig. 18. Based on Model-1, the optimal policy for maximizing the lysine production in a fed-batch culture during a fixed operation time was determined. The solution is similar to that for other microbial cultivations, such as in histidine and glutathione production as already explained. That is, the maximum production is attained by two production stages composed of an initial maximum cell growth phase followed by a maximum production phase. The optimal solution based on Model-1 was realized experimentally as shown in Fig. 19 and compared with the solution based on Model-2 by the computer simulation shown below.

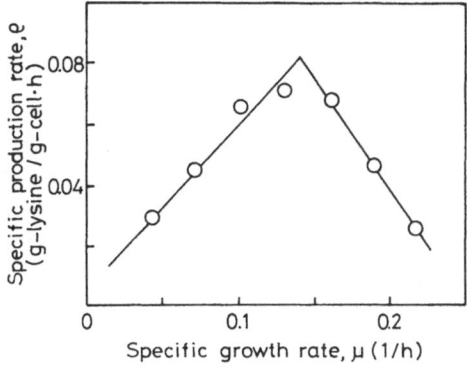

Fig. 18. Relationship between specific lysine production rate and specific growth rate

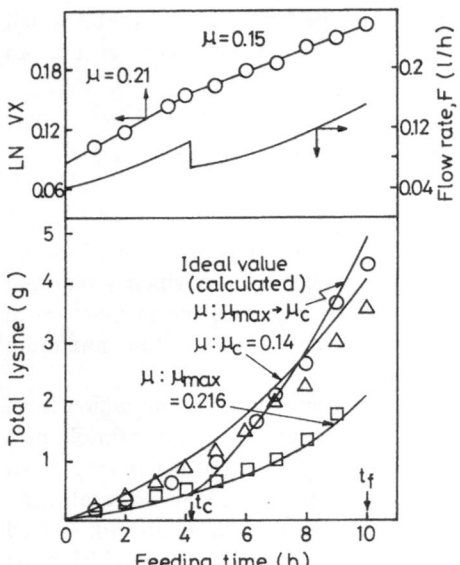

Fig. 19. Result of maximum lysine production in fed-batch culture

5.3 Comparison of Optimal Policies in Two Approaches

In a previous study (28), a mathematical model (Model-2) of lysine production was proposed. The mathematical model was built up based on the knowledge that leucine repressed lysine synthetic enzyme production and that the repression mechanism was regulated by the operator. The result is shown below.

$$\frac{dX}{dt} = \left(\mu - \frac{F}{V}\right) X \tag{51}$$

$$\frac{dS}{dt} = -v_G X + \frac{F}{V}(S_F - S) \tag{52}$$

$$\frac{dL}{dt} = -v_L X + \frac{F}{V}(L_F - L) \tag{53}$$

$$\frac{dP}{dt} = \varrho X - \frac{F}{V} P \tag{54}$$

$$\frac{dE}{dt} = K_M \mu - \mu E - K_D E \tag{55}$$

$$\frac{dV}{dt} = F, \tag{56}$$

where

$$\mu = \frac{\mu_{max}S}{K_S + S} \frac{L}{K_L + L}, \qquad v_G = m + \frac{1}{Y_{X/S}}\mu + \frac{1}{Y_{P/S}}\varrho$$

$$v_L = \frac{1}{Y_{X/L}}\mu, \qquad \varrho = \frac{\alpha_{max}S}{K_\alpha + S}E$$

$$K_M = A_1 + \frac{A_2}{1 + A_3L^n}.$$

Parameters for these equations are listed in Table 1. A comparison of the experimental relationship between μ and ϱ and the calculated one from Model-1 and Model-2 at the quasi-steady state is given in Fig. 20. The experimental data seem to fit Model-1 rather better than Model-2. However, for comparison, we assume here that Model-2 is correct and that Model-1 is an approximated model for Model-2. The problem considered here is to evaluate the difference between the optimal policies obtained based on both mathematical models. Also, the calculated objective functions are compared.

For Model-2, the optimization problem is described as; to find $F(t)$ for maximizing the amount of lysine at the final time T_f ($=$ fixed) subject to Eqs. (51) $\sim$ (56). For comparison of the optimal profiles calculated from both models, Model-1 and Model-2, at the same base, the inequality constraint for the

Table 1. Parameters for Model-2

$\mu_{max} = 0.216$	$\alpha_{max} = 0.85$	$K_S = 0.075$	$K_L = 0.0005$	$K_\alpha = 0.005$	
$Y_{X/S} = 0.21$	$Y_{P/S} = 0.41$	$Y_{X/L} = 17.2$	$m = 0.00001$	$n = 1$	
$K_D = 0.6$	$A_1 = 0.1$	$A_2 = 0.45$	$A_3 = 500.0$	$P_0 = 0$	
$X_0 = 1.757$	$V_0 = 1.32$	$S_0 = 41.3$	$L_0 = 0.001$	$E_0 = 0.044$	
$S_F = 50.0$	$L_F = 0.7$				

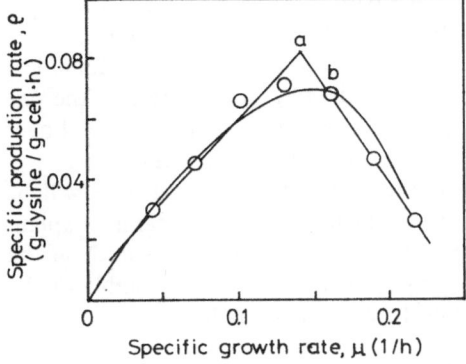

Fig. 20. Comparison of Model-1 *(straight line)* and Model-2 for prediction of ϱ from μ at a quasi-steady state. ○: experimental data

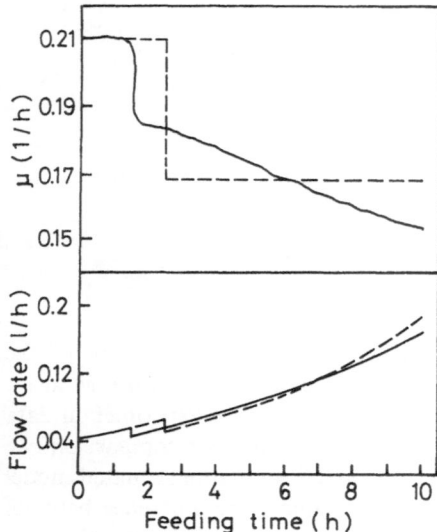

Fig. 21. Optimal policies calculated from Model-1 b[*] (– – –) and Model-2 (———)
[*]: ϱ_{max} is taken as pt. b in Fig. 20

manipulating variable $F(t)$ for Model-2 is given as follows:

$$\mu_{min} \leqq \mu(F(t)) \leqq \mu_{max}. \tag{57}$$

The control law, $F(t)$, is also evaluated as $\mu(t)$ instead of $F(t)$.

The optimal solution was obtained by direct searching for a nonlinear programming problem which was deduced from the original problem by the discretization of $F(t)$. The result is shown in Fig. 21. At can be seen from this figure, the optimal control policy was; at first, μ should be kept at μ_{max} and should be gradually decreased in just the same manner as already explained in the case of the nonlinear function of $\varrho(\mu)$.

When Model-2 is assumed to be correct, optimal policies based on each model were compared as shown in Table 2. As can be seen from this table, the objective function based on Model-1 is 5% less than that based on Model-2. If the maximum value of ϱ, ϱ_{max} is evaluated more accurately, e.g. taken as the value shown at the point b in Fig. 20, the objective function J becomes larger and is only 2.5% less than the true optimal value. In this case, the profile of μ is also close to the one from Model-2 as shown in Fig. 21. The most significant point is to be careful over the maximum value of ϱ, which will strongly affect the optimal policy.

From these results, it can be said that the approach employed here will be acceptable even if the model describing the specific production rate cannot explain the dynamic response with respect to μ. Of course, an approach based on the simplified model should be taken when a rigorous model is not available or the optimal solution will be difficult to obtain. In many fed-batch bioreactors, such situations can be expected to exist.

6 Concluding Remarks

In this article, an approach for the optimization of bioproduction and its realization in batch and fed-batch cultures was proposed. The approach has proven to be useful in many cultivation processes. However, there are still some difficult problems to be overcome, especially concerning the techniques of controlling the specific growth rate.

Of course, this approach will not be suitable for more complex processes in which the production rate cannot be correlated directly with the present value of the specific growth rate, such as in secondary metabolite production. For such cultivation processes, the history of the specific growth rate change during the reaction has the possibility of describing and controlling the production rate, and this may allow the approach proposed here to be applied successfully.

Table 2. Comparison of J

	J: total produced lysine (g)	ratio
Optimal Solution based on Model-2	3.82	100%
based on Model-1[a]	3.61	95%
based on Model-1[b]	3.73	98%

[a] ϱ_{max} is taken as pt. a in Fig. 20
[b] ϱ_{max} is taken as pt. b in Fig. 20

7 References

1. Porulekar SJ, Lim HC (1985) Modelling, optimization and control of semi-batch bioreactors. In: Fiechter A (ed) Advances in Biochemical Engineering, vol 32. Springer, Berlin Heidelberg New York, p 207
2. Modak JM, Lim HC. Tayeb YJ (1986) Biotechnol Bioeng 28: 1396
3. Lim HC, Tayeb YJ, Modak JM, Bonte P (1986) Biotechnol Bioeng 28: 1408
4. Yamane T, Shimizu S (1984) Fed-batch techniques in microbial processes. In: Fiechter A (ed) Advances in Biochemical Engineering, vol 30. Springer, Berlin Heidelberg New York p 145
5. Pontryagin LS, Boltyanskii VG, Gamkrelidze RV, Mischenko EF (1962) The mathematical theory of optimal processes (English translation by Trirogoff KN) Wiley-Interscience, New York
6. Aiba S, Nagai S, Nishizawa Y (1976) Biotechnol Bioeng 18: 1001
7. Dairaku K, Yamasaki Y, Kuku K, Shioya S, Takamatsu T (1981) Biotechnol Bioeng 23: 2069
8. Takamatsu T, Shioya S, Chikatani H, Dairaku K (1985) Chem Eng Science 40: 499
9. Takamatsu T, Shioya S, Okada Y, Kanda M (1985) Biotechnol Bioeng 27: 1675
10. Shimizu H, Shioya S, Suga K, Takamatsu T (1989) Appl Microbiol Biotechnol 30: 276
11. Shimizu H, Shioya S, Suga K, Takamatsu T (1989) Biotechnol Bioeng 33: 354
12. Dairaku K, Izumoto E, Morikawa H, Shioya S, Takamatsu T (1983) J of Ferm Technology 61: 189

13. Shioya S, Shimizu H, Ogata M, Takamatsu T (1985) Proc of 1st IFAC Symp on Modeling and Control of Biotechnological Processes, the Netherland, p 49
14. Jazwinski AH (1970) Stochastic processes and filtering theory. Academic Press, p 272
15. Stephanopoulos F, San KY (1984) Biotechnol Bioeng 26: 1176
16. Staniskis J, Simutis R (1986) Biotechnol Bioeng 28: 362
17. Caminal G, Laufente J, Lopez-Santin J, Poch M, Sola C (1987) Biotechnol Bioeng 29: 366
18. Nahilik J, Burianec Z (1988) Appl. Microbiol. Biotechnol. 28: 128
19. Wang HY, Coony CL, Wang DIC (1977) Biotechnol Bioeng 19: 69
20. Wang HY, Coony CL, Wang DIC (1979) Biotechnol Bioeng 21: 975
21. Montague GA, Morris AJ, Wright AR, Aynsley M, Ward A (1986) Canadian J of Chem Eng 64: 567
22. Grosz R, Stephanopoulos G, San KY (1984) Biotechnol Bioeng 26: 1198
23. Stephanopoulos G, San KY (1982) ACS Symposium Series 196: 155
24. Dochain D, Bastin G (1985) Proc of 1st IFAC Symp on Modeling and Control of Biotechnological Processes, the Netherlands, p 1
25. Chim-anage P, Shioya S, Suga K (1990) J Ferment Bioeng 70: 386
26. Chim-anage P, Shioya S, Suga K (1991) J Ferment Bioeng 71: 186
27. Shimizu H, Araki K, Shioya S, Suga K (1991) Biotechnol. Bioeng. 38: 196
28. Shioya S, Omasa T, Suga K (1988) Proc of 8th Int Biotechnology Symp Paris D: 154

Impacts of Automated Bioprocess Systems on Modern Biological Research

B. Sonnleitner, and A. Fiechter
Institute for Biotechnology, ETH Zürich Hönggerberg, CH-8093 Zürich, Switzerland

Bioprocess automation is mandatory to boost biotechnology from an empirical discipline to a natural science. Bioprocesses can be made more reproducible and predictable than generally expected by exploitation of modern high-tech biotechnology hard- and software.

Consequently, biological effects can be distinguished from technical ones, physiological states can be automatically detected and differentiated. Entire populations can even be forced into a certain, desired physiological state. Mathematical models help to resolve analytical insufficiencies and are most helpful in validating mechanistic hypotheses. With all the valuable information from hard- and software sensors available nowadays, the way is paved for the routine implementation of high performance process operation, for instance of integrated processes.

1 Introduction

It was only in 1985, that A. Johnson stated in his opening lecture to the 1st IFAC symposium on modelling and control of biotechnological processes that "bioprocesses are inherently unreproducible ..." and that this fact be "... reason enough to promote dynamic optimization and physiological control". No one in the audience had any objection to the first part of this statement at that time. However, in recent years, development and improvements of high performance

Advances in Biochemical Engineering/
Biotechnology, Vol. 46
Managing Editor: A. Fiechter
© Springer-Verlag Berlin Heidelberg 1992

bioreactors, on-line analytical facilities and application of automatic control to bioprocesses have brought evidence that this statement must be contradicted. In fact, biological systems do behave in highly reproducible and predictable manner as long as evolutionary changes can be neglected. This finding is of paramount importance for the future fate of (micro)biology because reproducibility of experiments, measurements and effects is an indispensable basis for all natural sciences. Fortunately, biotechnology has come forward in a successful way from a (mystic) art to a high technology; this evolution is nowadays recognized and honoured for instance by the fact that the EFB (European Federation of Biotechnology) has created and promoted the technical term "bioprocess sciences" which will, hopefully, replace the non-scientific and ambiguous term "fermentation". This is a visible sign of the tremendous changes that have started to take place in recent years.

In the present contribution, important aspect will be discussed and examplified by experiences with the eukaryote *Saccharomyces cerevisiae* which is of equal importance for scientific as well as industrial biological research.

2 Bioprocesses: Reproducible and Predictable

Reproducibility of bioprocesses is a prerequisite for success in both academia and industry, i.e. for sound biological research and high quality production. As a matter of fact, all organisms observe their own strategies of optimizing proliferation, protection and survival, however, they respond in different ways and to different degrees to certain changes of the environment. But the relevance of distinct conditions and sensitivity of the organisms' response are a priori not known. As a consequence of this ignorance, the worst case must be assumed and all conditions, as far as they can be influenced and affected by the process operator or computer, must be brought under control.

The usual way is to keep the environmental factors constant by dedicated closed control loops, i.e. make them so called culture parameters. This has been realized for many years for those variables that could be measured easily, e.g. temperature, pressure, or pH value. But obviously, these few physical and chemical variables are not the only ones determining the fate of proliferating and/or producing populations of microbes or cells. In recent years, more and more sensors and analytical devices for other variables have become available [1–4]. In fact, their consequent application and exploitation in bioprocesses has led to a considerable improvement of the environmental culture conditions and, as expected, to a higher reproducibility of cultivations. A survey of the capabilities of modern high-tech equipment is given in Table 1.

Unfortunately, biologically relevant state variables such as growth balance or imbalance, energy charge, load of storage materials or the dynamics of intracellularly accumulated substrates or products, degree of (de)repression, or the position in the cell cycle are even nowadays not yet directly accessible to online

Table 1. Characteristic ranges, relative accuracy of typical biotechnological state variable measurements and/or precision of cultivation parameters (i.e. control variables; according to [5])

Variable (analytical frequency)	Type/range	Units	Accuracy/precision (relativ to range)
temperature	0–150	°C	0.01%
rpm	0–3000	min^{-1}	0.2%
pressure	0–2	bar	0.1%
weight	90–100	kg	0.1%
	0–1	kg	0.01%
liquid flux	0–8	$m^3 h^{-1}$	1%
	0–2	$kg\, h^{-1}$	0.5%
dilution rate	0–1	h^{-1}	<0.5%
gas flux	0–2	vvm	0.1%
foam	on/off	—	—
bubbles	on/off	—	—
level	on/off	—	—
pH	2–12	units	0.1%
pO_2	0–100	% sat	1%
pCO_2	0–100	mbar	1%
exhaust-O_2	16–21	%	1%
exhaust-CO_2	0–5	%	1%
fluorescence	0–5	V	—
redox	−0.6–0.3	V	0.2%
MS: volatiles			membrane pervapo-
methanol, ethanol	0–10		ration properties
acetone	0–10	$g\, l^{-1}$	
butanols	0–10		1–5%
on-line FIA:			$= f$ (dilution)
glucose ($<100\, h^{-1}$)	0–100		<2%
NH_4^+ ($<'20\, h^{-1}$)	0–10	$g\, l^{-1}$	1%
PO_4''' ($<15\, h^{-1}$)	0–10		1–4%
GLU ($<20\, h^{-1}$)	0–10		<2%
on-line HPLC:			
phenols ($<5\, h^{-1}$)	0–100	$mg\, l^{-1}$	2–5%
phthalates	0–100	$mg\, l^{-1}$	2–5%
organic acids	0–1	$g\, l^{-1}$	1–4%
erythromycins	0–20	$g\, l^{-1}$	<8%
other by-products	0–5	$g\, l^{-1}$	2–5%
on-line GC			capillary columns
acetic acid ($<10\, h^{-1}$)	0–5		2–7%
acetoin	0–10		<2%
butanediol (R, S)	0–10	$g\, l^{-1}$	<8%
meso-butanediol	0–10		<8%
ethanol	0–5		2%
glycerol	0–1		<9%
RQ	0.5–20	$M\, M^{-1}$	largely depends on error propagation
x:			
OD-sensors	0–100	AU	highly variable
βugmeter	—		many interferences
physiological state	software	—	reference patterns
$\mu\ (= D)$	0–1	h^{-1}	<0.5%

analysis; and with honesty, it is mostly a vision with currently available off-line instruments, too. And this is why a residual uncertainty in both determination and control of a population's behavior still remains. However, this consequence will certainly be minimized in the future by improving frequency, selectivity and accuracy of measurements on the one hand and by expanding the spectrum of the relevant intra- and extracellular components on the other hand. The present state of the art with respect to quality and frequency of bioprocess analyses is paradigmatically characterized by Fig. 1 for a series of *Saccharomyces cerevisiae* cultivations.

There is no doubt that considerable improvements have already been realized and will certainly be extended in the near future. This expectation is realistic according to theoretical considerations: even in small, laboratory scale bioreactors there is a statistically relevant, high number of cells — easily exceeding 10^{10} by orders — contributing to what is observed macroscopically as growth and/or product formation. Individuality of distinct cells no longer has a significant effect

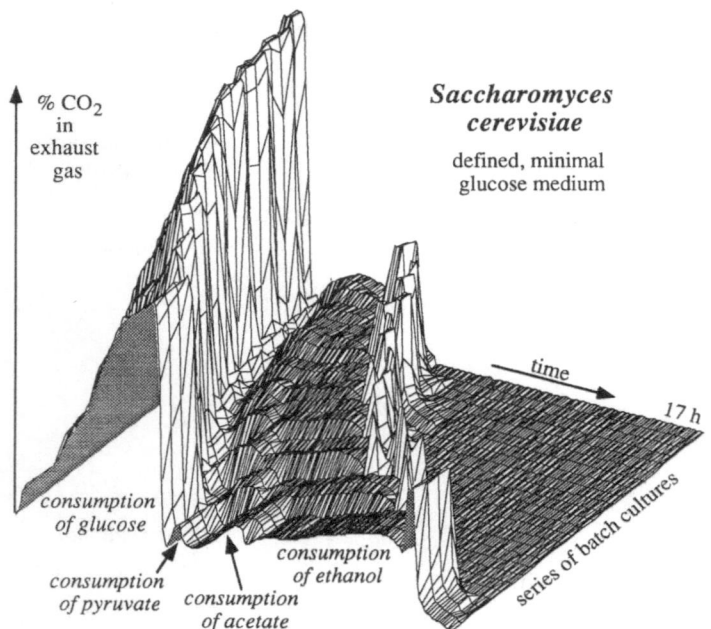

Fig. 1. Series of repetitive batch cultures of *Saccharomyces cerevisiae* with an increasing amount of inoculum left in the reactor.
From left to right: relative cultivation time, longest experiments: 17 h, time step between plotted data: 5 min.
From front to back: 1st batch culture inoculated from shake flask, repetition of experiment after harvesting of 95% of the culture and refilling with sterile medium, 2nd and 3rd repetition, next series with only 90% being replaced (i.e. 10% being left as inoculum), and so on until only 50% have been replaced (last series).
From bottom to top: CO_2 in the exhaust gas (scale: 0 to 5%)

in such populations and spontaneous mutants can be neglected, provided the cultivation conditions do not prefer them significantly and the observation period does not span evolutionary eras.

3 Distinction Between Biologically and Technically Caused Effects

In the past, biologists as well as engineers have basically attributed observed deviations from expected trajectories to the inherent variability of living systems.

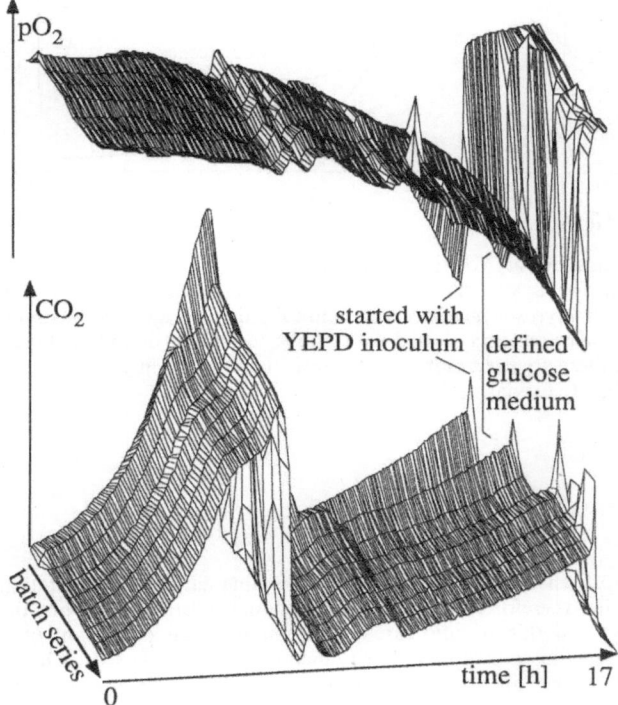

Fig. 2. Reproducibility demonstrated with a series of repetitive batch experiments (pO_2 and CO_2 in the exhaust gas shown):
A 1*st* batch in chemically defined glucose medium was inoculated with a shake flask precultue grown on complex medium (YEPD).
The 2*nd* batch was a continuation of the 1*st* after replacing 95% of the culture suspension with fresh defined medium, and so forth. In spite of uptake of some (unknown) components introduced with the complex medium and the dilution of these components due to liquid exchange the differences between 1*st*, 2*nd* and subsequent batches are significant, relevant and reproducible (no differences are observed when a shake flask inoculum grown on defined medium is used); they reflect a biological effect caused by some component from YEPD and definitely not an operational or technical effect. Culture: *S. cerevisiae*, 3% glucose medium, not oxygen limited, 30.0 °C, pH 5.0

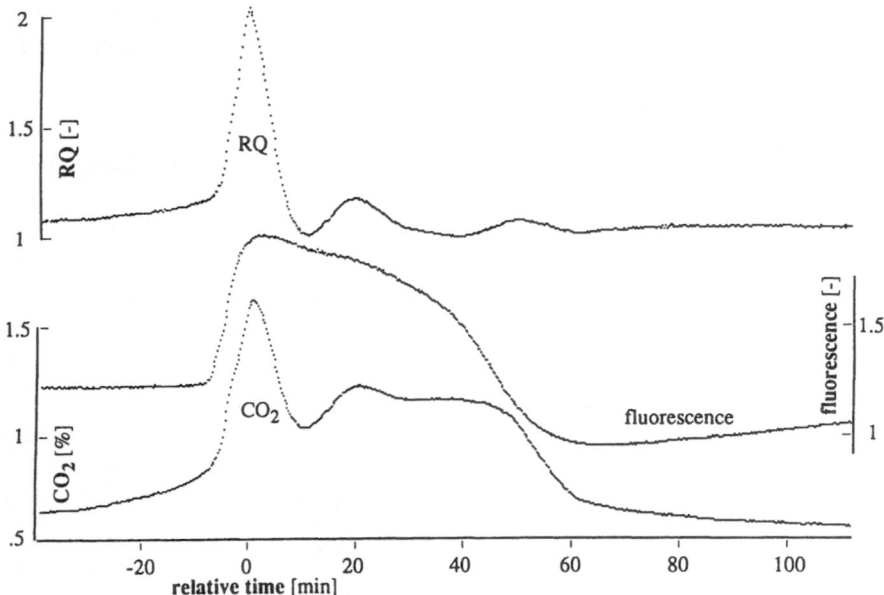

Fig. 3. Importance of high data density for the quantitative characterization of microbial culture dynamics visualized for a typical oscillation in spontaneously synchronizing continuous cultures of *Saccharomyces cerevisiae* on a chemically defined minimal medium. All 3 curves are continuous measurements plotted with a time resolution of 15 s (between consecutive dots). However, there are even more rapid transients observed in yeast cultures but less impressive to visualize (c.f. final peaks in Fig. 2)

Fig. 4. An example for comprehension of a huge amount of on-line data by interpretation ▶ as a "pattern": this is a finite, therefore restricted, 3-dimensional information about the proliferation of a population, in this specific case a fraction from an oscillation of a synchronized culture of *S. cerevisiae* grown on defined glucose medium at $D = 0.13 \, h^{-1}$. This graphic representation needed an arbitrary way of scaling which is − e.g. during the preparation of data for a pattern recognition algorithm − of course made according to defined rules, e.g. normalization of the covered range to the interval [0 1], possibly after spike elimination or a logarithmic transformation.

The *2nd* dimension (time) can, with advantage, be transformed into the frequency domain (Fourier transformation) in order to make the automated algorithm insensitive to temporal shifts (not shown for better visual perception).

Top: An entire reference pattern of the *1st* part of a synchronous oscillation derived from a series of individual oscillations by simple superposition of the respective data. In order to achieve a more detailed temporal characterization, this pattern would normally be divided in several subpatterns, e.g. 6 (c.f. [7]).

Bottom: A set of actually measured data from one distinct (= single) experiment covering a period of 30 min; this long period (only a fraction of data displayed) was selected for easier visual recognition and association with the respectively indicated subpattern of the reference

Thus, many interesting and even important paths to elucidate biological regulation and component ↔ function relationships were not followed. This kind of fatalism must be quickly overcome by exploiting the high performance systems available. There are presently 2 solutions visible which pave the ways to this objective:

1) Better reproducibility and reliability of cultivation techniques improves, statistically, the probability of correct and significant discrimination between different origins.
2) Increased frequency and accuracy of measurements also improve statistical safety and allow one to resolve so far ignored, overseen or misinterpreted effects. Figures 2 and 3 continue the above example in these respects.

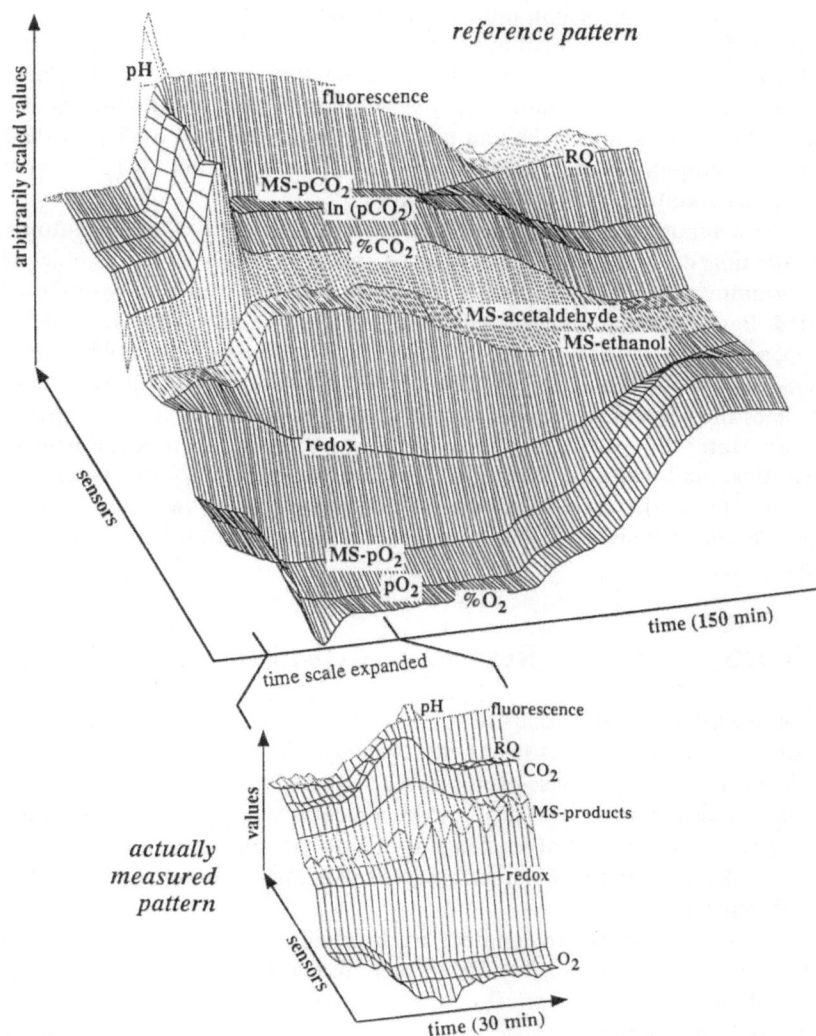

Fig. 4

4 Physiological States: Detection and Differentiation

A distinct physiological state can certainly not be defined by a single characteristic value or a single time trajectory. It is therefore important to collect and analyze measured or calculated (i.e. derived from raw) data in an at-least three dimensional way: every signal has an absolute value (which may be normalized by a reference value: 1st dimension). All these signals vary with the independent variable (usually time or dilution rate: 2nd dimension). However, there are generally several signals for different variables available in parallel (dependent state variables: 3rd dimension). Only this multivariate information describes a certain physiological state more or less clearly and sufficiently. The degree of uncertainty depends on density in the 2nd dimension, multiplicity in the 3rd dimension and on the size and position of the investigated window with respect to the 1st and 2nd dimension (i.e. distortion and shift) of the available process information. Projection of a huge amount of individual raw data into physiological states is a highly intelligent method of data reduction. A human user would normally tend to display the raw information graphically and to classify the corresponding pattern (e.g. Fig. 4) most probably by intuition [6, 7].

Process computers, however, allow us to perform this on-line (i.e. fully automatic) with little time delay (i.e. nearly in real time) on a traceable basis (i.e. as objectively as programmed) [6–9]. Possible algorithms are diverse and can be quite complicated. Pattern recognition algorithms try to find the best association between a given, actual pattern and a finite set of reference patterns which may be retrieved from a historical data base of from model simulations [7]. Fuzzy reasoning overcomes the uncertainty of association of an actual state vector or matrix with idealized states by calculating so called membership functions. Fortunately, statistical methods can give an estimate for the probability of a correct association. This estimate can be used to enable automatic control in the case of the most likely safe conclusions only and so effectively prevent doubtful decisions from being executed.

5 Models: A Help to Resolve Analytical Insufficiencies

Part of scientific research consists of the generation of hypotheses and their subsequent experimental testing. Unfortunately, the latter sometimes ends in non-significant results because of still insufficient analytical resolution and power. Though this situation is not typical for biological sciences the complex nature of living systems poses a special challenge to the development of analytical systems. Since this takes time, indirect methods must be used and exploited today; modelling is an efficient one.

It is helpful in the present context to distinguish 3 classes of models: those simple models used in every day life to such an extent that the model character is normally forgotten, e.g., optical density of a culture is expected to be linearly correlated with either cell *number* or *mass* concentration. Another class of models, those structurizing cells and segregating populations, is an important tool in molecular biology and physiology in order to investigate and validate molecular mechanisms. These models are most often so complex and tedious to calculate that supercomputing power is nowadays needed to succeed in reasonable time. The third class in this context is a manageable compromise aiming at the on-line use on process computers (c.f. Fig. 5). Simple origin ↔ effect mechanisms or even simpler algebraic functions substitute for the underlying molecular mechanisms resulting in models which can be repeatedly and quickly calculated on-line but with a restricted use for predictions outside the data space exploited for their creation [9]. Nevertheless, they are often a great help.

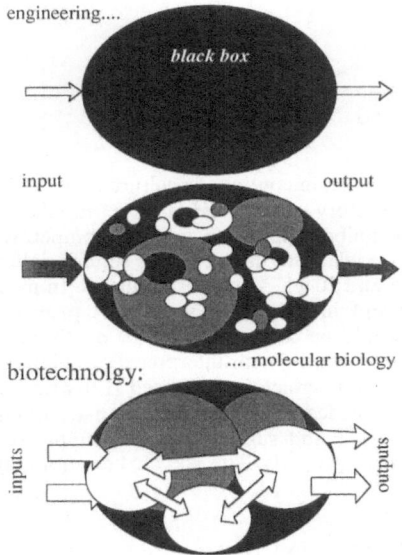

Fig. 5. Schematic comparison of different types of models (according to [5]).
Top: black box model widely used in engineering sciences: the biological systems has 2 connections to the environment: input and output. The characteristic property of the system is to convert the input vector according to rate equations into the output vector. The rate equations are of pure descriptive nature and must not be used to predict values in the space outside the original data base used to evaluate the model parameters.
Middle: collection of single, detailed informations about elements or subsystems of cells, typically the compiled knowledge accomplished in molecular biological investigations. Many interconnections (as rate equations) are normally ill defined or not known. Some structural elements are well understood and quantified (indicated by white areas), others less (grey zones) and the rest not yet (black).
Bottom: a compromise approach yielding an appropriate tool to predict (= calculate) reaction rates and yields. Biological knowledge is condensed to several relevant key elements. Their interconnection is defined by a relatively small but complete set of rate equations. Outputs are calculated from inputs but the algorithms are based on mechanistic assumptions; the mechanisms need not necessarily be molecular, even hypothetical mechanisms are helpful. Such models may — with the necessary criticism and care — be used to extrapolate outside the data space which served for construction and validation of the model

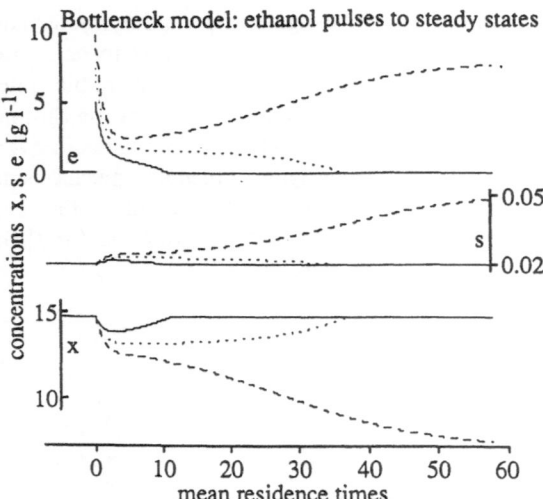

Fig. 6. Simulation of the response of bakers yeast grown in continuous culture close to the dilution rate D_R (where the saturation of the respiratory capacity takes place) to ethanol pulses at different levels. Ethanol is assumed to inhibit respiration in a noncompetitive manner with $K_i = 46 \text{ g l}^{-1}$. It is worth noting the very flat time courses of the state variables after more than 5 mean residence times for the 7.5 and 10 g l^{-1} pulse experiments. In most experimental studies not accompanied by modeling and simulation, this would most probably be mistaken as a steady state, experimentation would be continued after 10 or 20 mean residence times and the entire set of results would consequently be misinterpreted. Simulations were made according to the bottleneck model by Sonnleitner and Käppeli [12] with the parameter set given there (i.e. $D_R = 0.30 \text{ h}^{-1}$) just extended by ethanol inhibition according to Käppeli et al. [13] for $D = 0.19 \text{ h}^{-1}$, $s_0 = 30 \text{ g l}^{-1}$ and sufficient oxygen supply (no transfer limitation). The bifurcation is a model intrinsic property. This specific behavior may require sound process control (see [11])

It is very common that relaxation times of cultivation systems are not correctly anticipated even by experienced biologists [10]. Modelling brings necessarily quantitative aspects into biological reasoning and corrects efficiently human intuition, which is quite limited. Figure 6, for instance, gives an example for an intuitively unexpected behavior of bakers yeast in chemostat: depending on the extent of the disturbance (an ethanol pulse), the culture approaches two different steady states. However, they are reached only after a considerable time which makes the experimental proof very laborious [11].

6 Insight into Cells and Populations (Structurization and Segregation)

Modern biotechnology must exploit the understanding of molecular mechanisms. Their investigation has so far been successfully based on a series of invasive techniques, i.e. cells have been produced, disrupted by an appropriate method and more or less purified fractions have been studied. These in vitro experiments have

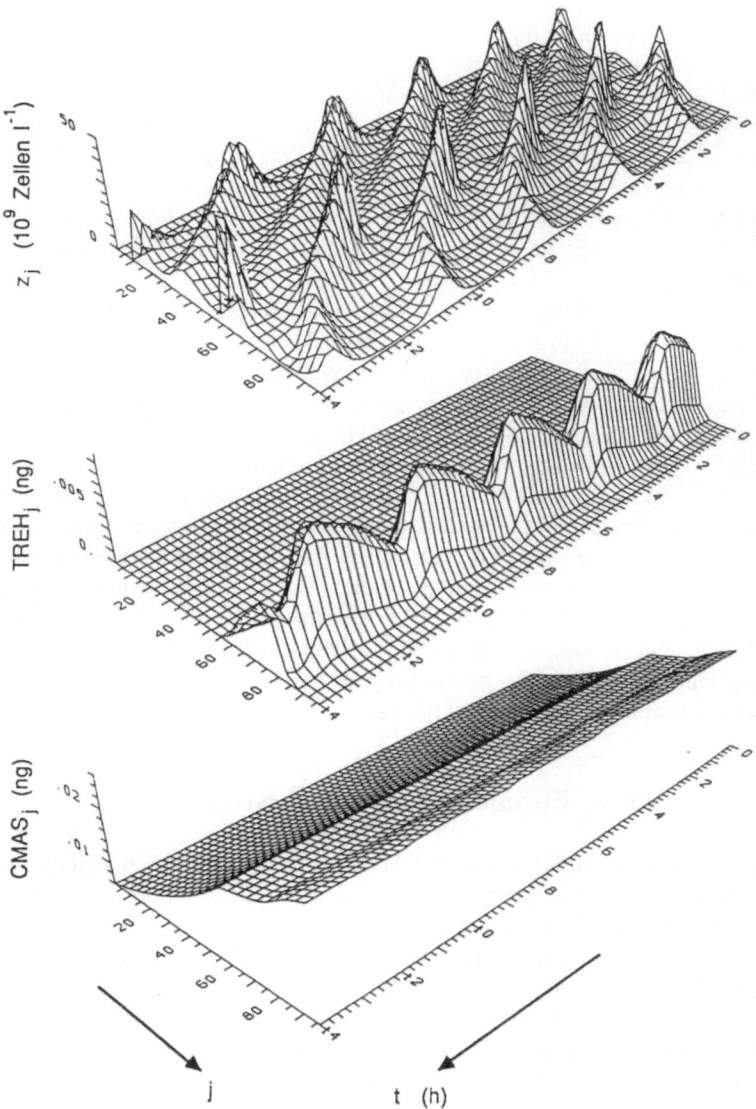

Fig. 7. Structured and segregated mathematical model for the proliferating biosystem *S. cerevisiae*. The dynamic solution was calculated for 97 cell classes (j = class number: 1 = smallest baby cells, 97 = double cells just prior to cell separation). Three structural elements were used to decribe the metabolism: respiratory bottleneck, storage carbohydrates and the residual biomass. Cell concentration (Z_j) is shown on top, the specific content of storage carbohydrates ($TREH_j$; per cell) in the middle and the residual biomass ($CMAS_j$; mass of individual cell) on bottom (from [14])

shed light on a great diversity of structural units in the cells and on their functions giving rise to the current molecular biological perception of life. Even a kind of dynamic pattern description can be achieved by the repeated performance of such a protocol. But this methodology suffers from an important constraint: a small subsystem is investigated which is isolated from a much larger and more complex entity, the whole cell or even the whole population. This simplification certainly obscures many functional and regulatory mechanisms that are decisive in vivo. As a consequence, noninvasive techniques must be applied to reveal these in − ideally − undisturbed systems, i.e. in well controlled cultures or well characterized ecological habitats.

Two complementary methodologies show promise of reaching this objective: noninvasive analytical techniques exploiting optical, electrical and electromagnetical principles (such as fluorometry, nuclear magnetic resonance, or infrared spectroscopy), and mathematical modelling of mechanistic hypotheses. They must complement each other in order to compensate the inherent weaknesses. But this requires perfect synchronization of hardware and software. The respective developments are becoming visible in reality: intelligent analytical subsystems have been coupled to bioreactors as well as process computers which are needed to evaluate and interpret large sets of representative data derived from noninvasive measurements. Inspite of the present infancy of these techniques, there is successful progress in getting dynamic insight into cells and entire populations (Fig. 7). This is manifested in structured and/or segregated mathematical models for proliferating biosystems which can nowadays be increasingly verified or even discriminated by reliable experimental measurements [14, 15].

7 Populations can be Forced into Desired States

The performance of a microbial or cellular population depends often strictly on its physiological state. This is a reason why special modes of operation have been established, e.g. the fed batch or Zulaufverfahren. However, such processes must often be operated in an open loop control mode because direct analytical measures for the desired physiological state are missing or there is too much delay. It is therefore important to apply or develop sufficiently rapid methods for the reliable estimation of states which is usually a combination of a set of measured (indirect) data and a predictive model (see other contributions in this issue). The key to success is to exploit highly selective and sensitive measures for all relevant variables.

For instance, an increase of culture fluoresence can be indicative for both an increase of biomass and/or an increase of a reductive pathway's activity. The first alternative, however, is slow whereas the second one is potentially fast. Evaluation of the time differential of this signal in combination with additional information about the gas exchange rates permits reliable discrimination between both causes and, consequently, a correct decision for required control actions (e.g. feed rate of limiting substrate; c.f. an example for *S. cerevisiae* by Meyer and Beyeler [16]).

Other aspects of cell proliferation or product formation are more complicated in that they are distinct transient events. There is basic interest in studying the

CPR [mole l⁻¹h⁻¹]   Fm [ml h⁻¹]

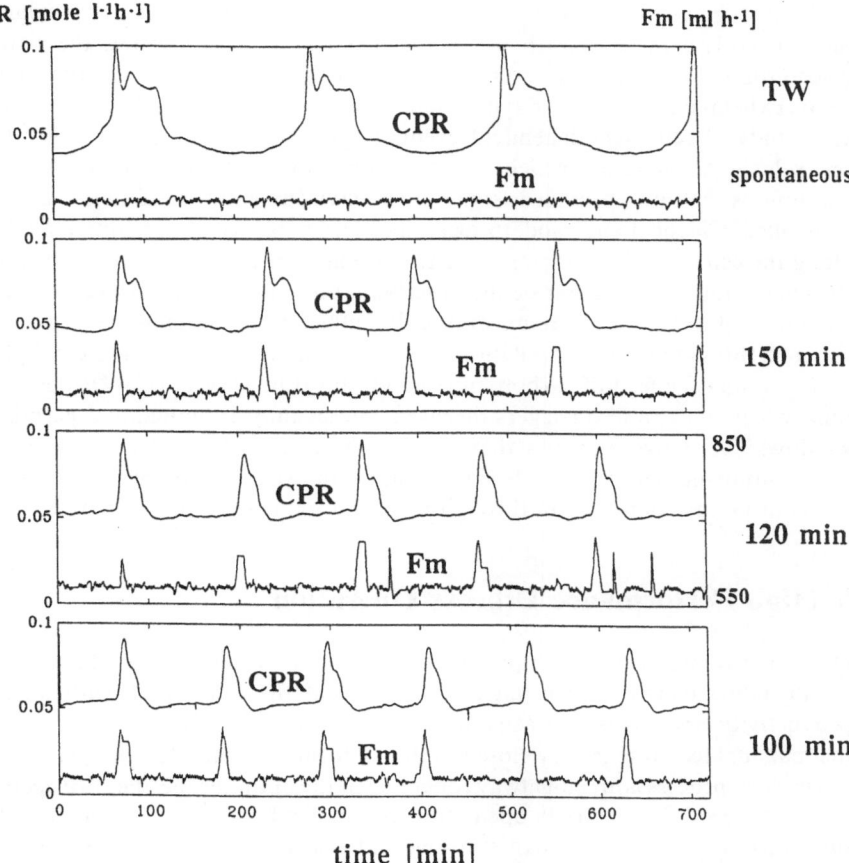

time [min]

Fig. 8. The time required by *S. cerevisiae* to proceed through the cell cycle can be forced to be less than under undisturbed culture conditions where the population shows spontaneous synchronization: the CO_2 production rate (CPR signal calculated from raw data) indicates the entry of a subpopulation into the S-phase of the cell cycle in a very sensitive way (sharp peak). When the flow rate of medium (Fm) is held constant, the mean generation time is 435 min under the selected operating conditions (time distance between) every *2nd* peak; top subplot).

If, however, the flow rate is increased for less than 10% during approximately 10 min, the subpopulation consisting of mature mother cells is forced to enter the S-phase earlier resulting in a shorter mean generation time (lower subplots). Reducing the time interval between flow rate peaks even further was found to be unsuccessful in forcing synchronous oscillations to higher frequency.

The conclusion is that, besides the extracellular trigger, namely exta supply of carbon and energy source through feed rate increases, the degree of maturity of cells is a determinant factor in enabling them to proceed through the cell cycle event sequence (present understanding of 'maturity' is: more than minimal cell size and amount of storage carbohydrates accumulated in the cells). Hence, the permissive window for possible generation times lies in the interval [200 435] min, where the upper limit is determined by an intracellular trigger (most probably: maximal load of storage materials reached; from [17]) chain of such events during the true cell cycle of eukaryotic cells. During balanced or constantly limited growth, i.e. batch, fed batch or chemostat, the natural distribution of mean generation times of cells within a population is large enough to overwhelm any phasing or synchronization effects. However, this makes the in vivo study of cell cycle dependent events impossible. An exception rather than the rule is *S. cerevisiae* which can be grown in time-invariant synchronized continuous culture, provided some microenvironmental conditions are well controlled. Classical and modern techniques allow plenty of descriptive analyses along the cell cycle. The findings give rise to many mechanistic hypotheses about the causes for observed transients. Discrimination and proof, however, can only be derived if these events are unequivocally influenced by forcing the population. Figure 8 shows that a subpopulation of *S. cerevisiae* can be forced into the S-phase by increasing the feed of carbon and energy substrate shortly [17]. However, the time, when an extra feed triggers this event successfully, is limited to a permissive window. This range indicates that the degree of maturity of the subpopulation representing mother cells is − besides an intra- or extracellular trigger − a typical prerequisite for continuation through the event sequence of the cell cycle.

8 High Performance Process Operation

Operating points which are optimal with respect to volumetric productivity are usually labile in praxis; for instance, in a chemostat, the dilution rate with maximal productivity is very close to wash out, or in a typical baker's yeast fed batch process the maximal feed rate is very close to ethanol formation and declining yield. This is why such processes are not always operated at the optimum, i.e. they are operated on the "safe side", when only open loop control can be applied. Although it often suffers from a lack of the necessary relevant real time measurements, the more professional solution is the implementation of closed loop control.

However, there is another improvement to this approach, namely the use of cell recycle systems. This operating principle also allows one to reach higher volumetric productivity due to increased biomass (= biocatalyst) concentration and to adjust the specific growth rate μ to the appropriate needs because, under steady state conditions, $\mu = D(1 − R)$, where R is the recycling ratio and D the dilution rate. Such integrated processes are, though mechanically more complex, easier to stabilize in praxis.

The example shown in Fig. 9 represents an interesting optimization problem with a combined objective function: the (optical) purity of the product is, along with the volumetric productivity of this biotransformation, of decisive importance. Both dilution rate and recycle ratio have been under closed loop control in order to force the specific growth rate to the physiologically most appropriate value; measurement of culture fluorescence and mass spectrometric detection of ethanol served to ensure fully oxidative metabolism of the yeast populations [18, 19]. Membranes have also been applied with great advantage in order to cope with extensive foam formation, e.g. in the case of biosurfactant production. The specialized membrane acted as a foam barrier and mediated gas transfer [20].

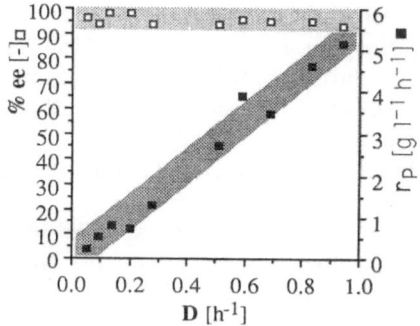

Fig. 9. Improvement of quality (enantiomeric purity, ee) and quantity (r_P). The volumetric productivity (r_P) of a culture can be significantly increased by recycling cells (e.g. with a cross flow filtration module): although the specific rates for growth (μ) and product formation (q_P) are kept constant, biomass increases with dilution rate and consequently also productivity. It is important to note that the optical purity (% ee) of the formed product (3-(S)-hydroxybutanoic acid ethyl ester) remains constantly high only when the physiological conditions − $\mu \approx 0.05\,h^{-1}$, pure oxidative metabolism, pH ≈ 2.2 − are strictly controlled (steady state). The leftmost data points (at $D = \mu = 0.05\,h^{-1}$) are representative for a single stage chemostat without cell recycling (according to [19])

9 Strategic Design of Cells and Bioprocesses

Future bioprocesses will differ considerably from the currently established "fermentations" in at least 3 major aspects:

1) The biological systems selected (i.e. the microbes, cells) will be physiologically well understood. The robust top performance hosts (TPH) will express the cloned genetic and regulatory information for the desired product(s) stably and efficiently. A most benefitting host property to be exploited will be their easy and reproducible cultivation.

2) All the relevant culture *variables* will be reliably measured and the identified, intentious culture *parameters* (among the set of variables) will be under precise closed loop control. The understanding of variables and parameters will certainly be expanded beyond the physical and the few chemical ones being considered at present and include the most decisive biological ones.

3) Production cultures will be supplied with the substances actually needed in the amounts actually required to accomplish a desired performance. Media and additives will possibly be mixed in situ from chemically defined components rather than using cheap complex materials which obscure the regulatory behavior and obstruct the down stream processing requirements.

The schedule to achieve these goals must comprise:

− a marked development of non-invasive analytical tools, sensors as well as autonomously intelligent analytical subsystems which permit one to analyze structural and/or functional entities within either the cells or entire populations (→ structured and segregated view of biocatalysts). Reliable interfaces to monoseptic processes must serve to connect further off-line analytics; off-line techniques will not become obsolete a priori, especially if sampling and sample preparation are better controlled and projectable.

— implementation of a larger number of dedicated automatic closed control loops with improved precision.

— development and improvement of (but also adaptation to specific bio-technological needs of elsewhere available) software to cope with the large amount of on-line data efficiently in real time (if this is necessary).
   This means:
   a) checking for plausibility or even validate the generated data,
   b) evaluating the plausible data with respect to their biological meaning resulting in efficient data reduction while avoiding information loss,
   c) comparing actual state matrices with reference data, to find associations and transforming them to decision making algorithms. The reference data may be either a conditioned experimental (= historical) data set or the results of simulations of mathematical models for biotechnological culture systems (i.e. including both biological and reactor part),
   d) developing, extending and improving fully automated methods of decision making for process control which must be both safe and secure; fault tolerance and redundancy will be further important features.

— automation of process optimization, for instance with regard to (possibly time variant) medium composition and culture conditions. Mechanistic models are expected to serve this objective on further improvements but heuristic (numerical) methods perform as well provided the penalty function can be reliably evaluated on-line.

— creation of a comprehensive framework to interpret measured data with respect to physiology, regulation and the evaluation of the metabolic network's rigidity. These aspects are of paramount importance for metabolic engineering basics and for production runs as well.

The tight integration of basic biological research with analytics and automatic control, material sciences and engineering sciences is a promising way to *make the biological systems do what biotechnologists intend them to do.*

10 References

1. Schügerl K (1990) ECB5, Copenhagen, Proc 2: 1010
2. Scheper T (1991) Bioanalytik. Vieweg, Braunschweig, D
3. Locher G, Sonnleitner B, Fiechter A (1992) J Biotechnol (in press)
4. Schügerl K, Lübbert A, Scheper T (1987) Chem-Ing-Tech 59: 701
5. Sonnleitner B (1991) Ant v Leeuwenhoek 60: 133
6. Locher G, Sonnleitner B, Fiechter A (1991) J Biotechnol 19: 127; 19: 173
7. Locher G, Sonnleitner B, Fiechter A (1990) Bioproc Eng 5: 181
8. Dors M, Behrendt J, Kreibaum U, Havlik I, Wingelsdorf R, Lübbert A (1991) Biotech Forum Europe 8: 612

9. Sonnleitner B (1989) Habilitationsschrift, ETH Zürich
10. Sonnleitner B (1991) Bioproc Eng 6: 187
11. Axelsson JP, Andersen MY, Jörgensen SB (1992) J Biotechnol (in preparation)
12. Sonnleitner B, Käppeli O (1986) Biotechnol Bioeng 28: 927
13. Käppeli O, Lorencez Gonzales I, Kühne A, Sonnleitner B (1988) 8th Int Biotechnol Symp, Paris, Proc 1: 467
14. Strässle C, Sonnleitner B, Fiechter A (1988) J Biotechnol 7: 299
15. Strässle C, Sonnleitner B, Fiechter A (1989) J Biotechnol 9: 191
16. Meyer C, Beyeler W (1984) Biotechnol Bioeng 26: 916
17. Münch T, Sonnleitner B, Fiechter A (1992) J Biotechnol 22: 329
18. Rohner M, Münch T, Sonnleitner B, Fiechter A (1990) Biocatalysis 3: 37
19. Rohner M, Sonnleitner B, Fiechter A (1991) J Biotechnol 22: 129
20. Gruber M (1991) PhD thesis, University of Stuttgart, Germany

Use and Engineering Aspects of Immobilized Cells in Biotechnology

S. Furusaki and M. Seki

Department of Chemical Engineering, Faculty of Engineering, The University of Tokyo, 7-3-1, Hongo, Bunkyo-ku, Tokyo 113, Japan

A short review of the research in the past two years (1990–1991) on immobilized whole cells, such as microbial, plant, and animal cells, is presented including a discussion from an engineering point of view. Recent works concerning the intraparticle mass transfer effect on immobilized microbial cells by the authors and their co-workers are also introduced. Finally, future prospects of the immobilized cell system will be discussed.

Advances in Biochemical Engineering
Biotechnology, Vol. 46
Managing Editor: A. Fiechter
© Springer-Verlag Berlin Heidelberg 1992

List of Symbols

D_{ep} = effective diffusivity of ethanol inside the gel particles $[m^2 h^{-1}]$
D_{es} = effective diffusivity of glucose inside the gel particles $[m^2 h^{-1}]$
D_{0p} = diffusivity of ethanol in water $[m^2 h^{-1}]$
D_{0s} = diffusivity of glucose in water $[m^2 h^{-1}]$
$I_p(P)$ = product inhibition function for cell growth $[-]$
$I'_p(P)$ = product inhibition function for ethanol production $[-]$
$I_s(S)$ = substrate inhibition function for cell growth $[-]$
$I'_s(S)$ = substrate inhibition function for ethanol production $[-]$
K_i = substrate inhibition constant for ethanol production $[g\,l^{-1}]$
K_p = product inhibition constant for cell growth $[g\,l^{-1}]$
K'_p = product inhibition constant for ethanol production $[g\,l^{-1}]$
K_s = substrate limitation constant for cell growth $[g\,l^{-1}]$
K'_s = substrate limitation constant for ethanol production $[g\,l^{-1}]$
k_1 = constant in the diffusivity relation $[-]$
k_2 = constant in the diffusivity relation $[l\,g^{-1}]$
P = product (ethanol) concentration $[g\,l^{-1}]$
P_{max} = maximum product concentration for cell growth $[g\,l^{-1}]$
P'_{max} = maximum product concentration for ethanol production $[g\,l^{-1}]$
P_i = transition concentration for product inhibition for cell growth $[g\,l^{-1}]$
P'_i = transition concentration for product inhibition for ethanol production $[g\,l^{-1}]$
r = distance from the center of particles $[m]$
r_p = specific production rate [g EtOH per g dry cell per h]
r_{p0} = specific production rate without product inhibition [g EtOH per g dry cell per h]
r_{pm} = maximum specific production rate [g EtOH per g dry cell per h]
r_s = production rate per unit surface area of cell-immobilizing particles [g EtOH per mm² per h]
R = radius of immobilized particle $[m]$
S = substrate (glucose) concentration $[g\,l^{-1}]$
S_{max} = maximum substrate concentration $[g\,l^{-1}]$
S_i = threshold concentration for substrate inhibition $[g\,l^{-1}]$
S_0 = substrate concentration in the bulk solution $[g\,l^{-1}]$
t = time $[h]$
X = cell concentration [g dry cell per l]
X_0 = initial cell concentration [g dry cell per l]
$Y_{p/s}$ = yield of ethanol from glucose [g EtOH per g glucose]
μ = specific growth rate $[h^{-1}]$
μ_0 = specific growth rate without product inhibition $[h^{-1}]$

1 Introduction

A great number of biological processes using various immobilized biocatalysts, such as enzymes, microorganisms, organelles, plant cells and animal cells, have been investigated for utilization of the advantages of immobilized biocatalysts. Increasingly, extensive applications of immobilized biocatalysts have been proposed since the first successful industrialization of the immobilized enzyme process by Chibata et al. of the Tanabe Seiyaku Co. in 1969.

Among various immobilized biocatalyst systems, immobilized whole cell systems, in particular, have received considerable attention. Advantages of whole (living) cell systems include their catalyzing ability to synthesize various useful and complicated chemicals using multi-enzyme steps, and regeneration activity to prolong their catalytic life. In contrast to ordinary suspension culture systems, immobilized whole cells have the merits of: (1) avoiding wash-out of cells at a high dilution rate; (2) higher cell concentration in the reactor; and (3) easy separation of cells from reactors or the product-containing solution. Due to their importance, engineering research of immobilized cells has also been realized and has covered a wide range of aspects, such as immobilizing materials, mass transport effects, physical and chemical environment, kinetics, and process modeling.

Some useful comprehensive reviews and many survey papers on several specific fields of immobilized cell applications have already been published [1-10]. In this paper, a short review of research on immobilized whole cells mainly in the past two years (1990-1991) is presented as a extension of those published reviews including the discussion from an engineering point of view. Next, recent works in the intraparticle mass transfer effect on immobilized microbial cells by the authors and their co-workers are presented. Finally, future prospects on the immobilized cell system will be discussed.

2 Immobilized Cells: A Mini-Review of Recent Literature

2.1 Immobilized Microbial Cells

2.1.1 Production of Alcohols

Various alcohols such as ethanol, butanol, isopropanol are produced from carbohydrates using immobilized whole cell systems. Among them, large-scale industrial ethanol production is already beyond the stage of pilot plant operation. However, its economic feasibility still depends on the oil market. A considerable amount of research has been carried out on ethanol production processes using immobilized microorganisms as model systems for immobilized whole cells.

Recent research on the production of alcohols using immobilized microorganisms is shown in Table 1. Although various immobilization techniques have been applied to ethanol production systems, most of the research has still adopted calcium alginate entrapment due to its simplicity and the mild conditions of the immobilizing procedure [11, 17-18, 21-24, 26-27]. However, immobilization

Table 1. Production of alcohols by immobilized microbial cells

Alcohol	Substrate	Support material	Microorganisms	Ref.
Ethanol	Sucrose	Ca-alginate	*Saccharomyces cerevisiae*	[11]
Ethanol	Sucrose	Alminosilicate	*Saccharomyces cerevisiae*	[12]
Ethanol	Glucose	Alginate	*Saccharomyces cerevisiae*	[13]
Ethanol	Lactose	Porous sponge	*Kluyveromyces fragilis*	[14, 19, 20]
Ethanol	Corn starch	Polethyleneimine-alginate	*Saccharomyces cerevisiae* (+Amylase)	[15]
Ethanol	Glucose	Polyurethane foam	*Zymomonas mobilis*	[16]
Ethanol	Glucose	Ca-alginate	*Saccharomyces cerevisiae*	[17]
Ethanol	Glucose	Ca-alginate	*Saccharomyces cerevisiae*	[18]
Ethanol	Glucose	Ca-alginate	*Zymomonas mobilis*	[21]
Ethanol	Glucose	Ca-alginate	*Zymomonas mobilis*	[22]
Ethanol	Glucose	Ca-alginate	*Saccharomyces cerevisiae*	[23, 24]
Ethanol	Glucose	Glass fiber mat	*Saccharomyces uvarum*	[25]
Ethanol	Glucose	Ca-alginate, Sintered glass	*Saccharomyces cerevisiae*	[26]
Ethanol, 4-ethyl-guaiacol	Glucose	Ca-alginate	*Zygosaccharomyces rouxii, Candida versatilis*	[27]
Butanol	Glucose	Ca-alginate	*Clostridium beyerinckii*	[28]
Butanol	Glucose	Polyester sponge	*Clostridium acetobutylicum*	[29]
ABE	Lactose	Bonechar	*Clostridium acetobutylicum*	[30]
Glycerol	Sucrose	Sintered glass	*Saccharomyces cerevisiae*	[31]
Polyols	Glucose	Sintered glass	*Pichia farinosa*	[32, 33]

ABE = acetone, butanol and ethanol

techniques other than gel entrapment of cells, which include adhesion and passive immobilization in porous materials have also been used for alcohol production.

A ceramic-like matrix material constructed of an aluminosilicate composite was packed in two glass columns and used for continuous ethanol production from synthetic sugar-cane juice containing $200 \, g \, l^{-1}$ sucrose [12]. The process was operated continuously for 2 years at $98 \, g \, l^{-1}$ of ethanol concentration. In this study, the productivity based on the reactor volume and ethanol yield was successfully maintained at high level for a long time, which seems to be competitive with the existing efficient processes. Amin and Doelle [16] used a polyurethane foam to entrap *Zymomonas mobilis* cells resulting in a high ethanol concentration at $120 \, g \, l^{-1}$. Chen et al. [14, 19–20] adopted a porous adsorbent sponge matrix as a supporting material. Immobilization using membranes, called a membrane bioreactor (MBR), has also been examined for ethanol production [34–35]. In comparing various systems of continuous ethanol production, it is important to consider all the factors affecting economic feasibility. Substrate concentration and its cost, product concentration and its yield, volumetric productivity, and long-term operational stability should be taken into account for process evaluation.

From an engineering point of view, long-term operational stability seems to be a salient problem to investigate. Dale et al. [19] considered long-term reactor performance taking the cell death-rate into account. Feeding strategies, in order to maintain minimal nutritional requirements for immobilized cells, were studied for long-term cultivation [20]. On the other hand, the transient dynamics of start-up performance for continuous ethanol production using immobilized cells has also been studied [21–22].

The immobilization effect on the metabolic pathway is one of the most interesting problems with respect to immobilization. Galazzo and Bailey [17] showed that non-growing *Saccharomyces cerevisiae* cells previously grown in alginate exhibit ethanol production rates 1.5 times greater than cells previously grown in suspension. Measurement of the metabolites concentration in the cells using in vivo nuclear magnetic resonance (NMR) spectroscopy suggested that the cells growing in alginate change their metabolic pathway at a genetic level. Hilge-Rotmann and Rehm [18] investigated the differences between free and calcium-alginate entrapped *Saccharomyces cerevisiae*. Increased specific hexokinase and phosphofructokinase activities could be determined in the cells. It was shown that these activations were connected with their growth characteristics forming aggregates in gel matrices. Further investigations will be needed to clarify these kinds of problems.

The contribution of immobilized cells and free cells released from gel beads to the production of ethanol and 4-ethylguaiacol (4-EG), which adds the characteristic aroma to soy sauce, was investigated [27]. In their results, ethanol and 4-EG were produced mainly by free cells and by immobilized cells, respectively. This phenomenon could be explained by product inhibition on growth.

In situ alcohol recovery using pervaporation was examined for the continuous production of isopropanol-butanol-ethanol or acetone-butanol-ethanol (ABE) with immobilized cells [28, 30]. Although pervaporation is thought to be a promising method for the in situ product recovery, membranes for this operation were frequently plugged by microbial cells [30]. Characterization of a trickle bed reactor during long-term ABE biosynthesis with immobilized cells was also investigated [29]. Co-immobilization of microbial membrane fragments to remove oxygen for the ABE biosynthesis was demonstrated [36].

2.1.2 Production of Organic Acids

Organic acids are extensively used in the food and pharmaceutical industries and some of them are products of microbial processes. Production of organic acids using immobilized microorganisms has also been investigated. In a process using acrylamide-gel immobilized dead cells of *Brevibacterium ammoniagenes*, industrial production of malic acid from fumaric acid has been carried out by the Tanabe Seiyaku Co. since 1974. However, in the cases of organic acid production using immobilized living cells, lactic acid has been investigated most extensively among various organic acids such as citric acid, itaconic acid, gluconic acid, acetic acid (Table 2). This is because the cultivation of lactic acid bacteria is little affected by the oxygen concentration, which could often be a limiting factor of a production system using immobilized cells.

Krischke et al. [37] carried out continuous production of L-lactic acid from whey permeate in a fluidized bed reactor with immobilized *Lactobacillus casei* cells bound on porous sintered glass beads. A volumetric productivity of lactate was $10 \text{ g l}^{-1} \text{ h}^{-1}$, which was a better result than that in a continuously stirred tank reactor.

Table 2. Production of organic acids by immobilized microbial cells

Organic acid	Substrate	Support material	Microorganisms	Ref.
Lactic acid	Lactose	Sintered glass	*Lactobacillus casei*	[37]
Lactic acid	Lactose	ϰ-Carrageenan Ca-alginate, Agar, Polyacrylamide	*Lactobacillus casei, Lactococcus lactis*	[38]
Lactic acid	Lactose	ϰ-Carrageenan-locust bean gum	*Streptococcus salivarius, Lactobacillus delbrueckii*	[39, 48]
Lactic acid	Glucose	ϰ-Carrageenan	*Lactobacillus delbrueckii*	[40]
Lactic acid	Raw starch	ϰ-Carrageenan	*Lactobacillus casei* (+Amylase)	[41]
Lactic acid	1,2-Propane-diol	Polyacrylamide Agarose Carrageenan	*Methylobacillus flagellatum*	[42]
Propionic acid	Glucose	Sintered glass	*Propionibacterium acidi-propionici*	[43]
Citric acid	Glucose	Ca-alginate, ϰ-Carageenan, Polyurethane gel, Nylon web, Polyurethane foam	*Yarrowia lipolytica*	[44]
Acrylic Acid	Propionate	Ca-alginate	*Clostridium propionicum*	[45]
Acetic acid	Ethanol	Ca-alginate	*Acetobacter aceti*	[46, 47]

Use of whey permeate or deproteinized whey, which is a waste product of the dairy industry, as a biosynthetic substrate will probably receive more attention due to the economic feasibility of the dairy by-products. Co-immobilization of *Lactobacillus casei* and *Lactococcus lactis* cells gave better results for the production of lactic acid from deproteinized whey than those of an immobilized single strain [38]. In this case, calcium alginate was the best immobilizing material compared with ϰ-carrageenan, agar, and polyacrylamide gels. Audet et al. [39, 48] also investigated a mixed culture of lactic acid bacteria, *Streptococcus salivarius* and *Lactobacillus delbrueckii*, immobilized separately in ϰ-carrageenan-locust bean gum gel beads. In this case, the increase of the lactate production rate by cooperation between the two strains was not observed.

For the alleviation of end product inhibition, extractive lactic acid production by immobilized *Lactobacillus delbrueckii* was investigated [40]. In the case of lactic acid production, it is important to determine the optimum product concentration, which yields an acid diluted enough to avoid the product inhibition while still being economically viable with respect to the purification process.

Using both amylase, immobilized on a carrier which could be reversibly soluble-insoluble depending on pH, and *Lactobacillus casei* entrapped in ϰ-carrageenan gel, lactic acid was continuously produced directly from raw starch at a volumetric productivity of $3.1 \text{ g l}^{-1} \text{ h}^{-1}$ [41].

Obligate methylotrophic bacterium cells, *Methylobacillus flagellatum*, are attractive as a means for lactic acid production from 1,2-propanediol, a cheap and available product of petrochemistry. Dinarieva and Netrusov [42] observed that the half-life of these cells immobilized in carrageenan was increased 5–6 times as compared with free cell suspensions.

The cells of aerobic bacterium *Acetobacter aceti* have been used for acetic acid production from ethanol. Continuous production of acetic acid using the immobilized cells of *A. aceti* in a three-phase fluidized bed reactor was investigated [46–47]. In this case, both immobilized and suspended cells contribute to the acetate production. Theoretical calculations showed that the optimal solid (gel) holdup and gel size existed at a high dilution rate.

2.1.3 Production of Antibiotics

Although about 150 antibiotics are now commercially produced, most of them are produced by microbial processes. One of the most important subjects related to antibiotic production using immobilized living cells is a continuous stable production of non-growth-associated secondary metabolites. Many applications have already been reported [9] (Table 3). One of the essential problems for

Table 3. Production of antibiotics by immobilized microbial cells

Antibiotics	Support material	Microorganism	Ref.
Candicidin	x-Carrageenan	*Streptomyces griseus*	[49]
Penicillin	Celite	*Penicillium chrysogenum*	[50]
Cyclosporin A	Porous celite	*Tolypocladium inflatum*	[51]
Nikkomycin	Sintered glass	*Streptomyces tendae*	[52, 53]

continuous, stable antibiotic production by immobilized microorganisms is how to regulate the concentration of growth-limiting substrates such as oxygen [52], nitrogen [53], phosphate [49, 53], and amino acid [51] to maintain biosynthetic activity of antibiotics. The periodic operation of a growth-limiting substrate feeding [49] is thought to be a good operational strategy. Keshavarz et al. [50] demonstrated continuous antibiotic production with low cell-growth rate using a cell-division cycle mutant which is temperature-sensitive.

2.1.4 Production of Enzymes

Enzymes are important industrial proteins suitable for microbial production. Hydrolytic enzyme production using immobilized cell systems has been investigated most extensively (Table 4).

Production of cellulase was investigated in a repeated batch culture system using immobilized mutants of *Trichoderma reesei* on a stainless steel mesh disk in a rotating disk reactor [54]. The enzyme production was maintained for three successive batch cycles (ca. 30 days) and the enzymatic activity was higher than that produced in a stirred tank reactor, since the low shear stress did not damage the enzyme.

Table 4. Production of enzymes by immobilized microbial cells

Enzyme	Support material	Microorganism	Ref.
Cellulose	Stainless steel mesh	*Trichoderma reesei*	[54]
Glucoamylase	Ca-alginate	*Aureobasidium pullulans*	[55, 56]
Glucoamylase	Sintered glass, Pumice stone	*Aspergillus niger*	[57]
Glucoamylase	Ca-alginate	*Aspergillus phoenicus*	[58]
α-Amylase, Pullulanase	Ca-Alginate	*Clostridium* sp., *Thermoanaerobacter* sp.	[59]
Lipase	Ca-alginate	*Sporotrichum thermophile*	[60]
Protease	Ca-alginate	*Myxococcus xanthus*	[61]
Protease	Ca-alginate	*Penicillium chrysogenum*	[62]
Lisosomal enzymes	Ca-alginate	*Tetrahymena thermophila*	[63]
Lignin peroxidase	Porous alumina	*Phanerochaete chrysosporium*	[64]
Lignin peroxidase	Polyurethane foam	*Phanerochaete chrysosporium*	[65]

Cells of *Aureobasidium pullulans* [55, 56] and *Aspergillus phoenicus* [58] were immobilized in calcium alginate gel beads and employed for continuous production of glucoamylase in a fluidized-bed reactor. Similarly to the production of antibiotics described above, separation of the growth phase and the production phase is effective in achieving higher enzyme productivity. Two-stage operation using two different media was examined for this purpose [58]. Sintered glass and pumice stone were also used as immobilizing materials for a repeated batch culture of glucoamylase production to obtain a higher accumulation of enzyme activity in the liquid medium [57].

Klingeberg et al. [59] investigated α-amylases and pullulanases production using immobilized anaerobic thermophillic bacteria that belong to the genera *Clostridium* and *Thermoanaerobacter*. Compared with free cells, the specific activity of the produced extracellular enzymes increased 5.6-fold.

Lipase production by immobilized protoplast of *Sporotrichum thermophile* was studied in batch culture [60]. Cell-wall-degrading enzymes and cellulase improved the lipase secretion of the normal mycelium by 25%–100%. It was shown that the specific lipase activity of immobilized protoplasts was about four times higher than that of normal mycelial beads.

The immobilization of *Tetrahymena thermophila*, a protozoan, was investigated for the production of lysosomal enzymes such as α-glucosidase, β-glucosidase, β-hexosaminidase and acid phosphatase [63]. These protozoa are used in diagnostics, and other analytical tests. Although the cells immobilized in ordinary calcium alginate gel beads were unable to grow, encapsulated cells in hollow calcium alginate spheres multiplied well.

Immobilization of lignin-degrading white-rot fungus *Phanerochaete chrysosporium* on polyurethane foam [65] and porous alumina spheres [64] was studied for the production of lignin peroxidase (ligninase). As these microorganisms have the potential to degrade the chlorinated lignins and low-MW chlorinated compounds which are the main water pollutants from kraft wood pulping, immobilization of these cells would contribute to the development of new processes to treat waste water. Kirkpatrick et al. [66] examined the immobilization of the white-rot fungus *Trametes versicolor* in polyurethane foam for biological bleaching of hardwood kraft pulp.

2.1.5 Biotransformation of Steroids

Much attention has been paid to the biotransformation of various steroids for the production of pharmaceutical steroid hormones. Since the steroid-transforming enzymes such as hydroxylase and dehydrogenase are considerably unstable, the use of immobilized microbial cells as biocatalysts has a significant advantage in steroid transformation.

The immobilization of microorganisms catalyzing various transforming reactions of steroids such as 11-β-hydroxylation, 11-α-hydroxylation, 9-α-hydroxylation, 16-α-hydroxylation, and Δ-1-dehydrogenation, has been investigated. Recent reports are summarized in Table 5.

Table 5. Transformation of steroids by immobilized microbial cells

Substrate	Reaction	Support material	Microorganism	Ref.
Cortisol	Δ-1-Dehydrogenation	Unwoven cloth coated with a copolymer	*Arthrobacter simplex*	[67]
Cortisol	Δ-1-Dehydrogenation	Polyacrylamide, Ca-alginate, Carrageenan	*Arthrobacter simplex*	[68]
Cortisol	Δ-1-Dehydrogenation	Ca-alginate	*Arthrobacter simplex*	[69]
4AD, ADD. Progesterone, Cortisol 16-dehydro-progesterone	Δ-4-Reduction	Polyacrylamide hydrazide	*Clostridium paraputricum*	[70]

2.1.6. Application to Waste Water Treatment

Waste water containing various pollutants such as cyanide, phenolics, ammonia, nitrates, and chlorinated compounds is usually treated by the activated sludge method. However, a large number of studies have been carried out on the effective removal of specific pollutants using microorganisms degrading specific compounds. Recent studies on these microorganisms are shown in Table 6. As waste water treatment is usually expected to achieve a large capacity and low operational costs, the investigation must be carried out on co-immobilization of a pair of microorganism strains or mixed culture systems for multiple degradations of various contaminants.

2.1.7 Other Applications

Some of the recent research on the production of other useful chemicals by immobilized microbial cells are shown in Table 7.

Table 6. Application of immobilized microbes to treat waste water

Substrate	Support	Microorganism	Ref.
Trimethyl lead	Polyacrylamide, Wood shavings	*Arthrobacter* sp., *Phaeolus schweinitzii*	[71]
3-Chloroaniline	Ca-alginate	*Pseudomonas acidovorans*	[72]
3,4-Dichloroaniline	Celite	Underfined	[73]
4-Chlorophenol	Granular clay	*Alcaligenes*	[74]
4-Chloro-2-phenol	Ca-alginate	*Enterobacter cloacae*, *Alcaligenes* sp.	[75]
Phenol	Activated carbon, Sintered glass	*Cryptococcus elinovii*, *Pseudomonas putida*	[76]
Nitrate	x-Carrageenan	Undefined	[77]
Ammonia	Agarose	*Thiosphaera pantotropha*	[78]

Some enzymatic reactions using microorganisms proceed efficiently in the presence of organic media. In these cases, immobilization of microbial cells will offer improved stability against the organic solvents. Immobilization of *Nocardia corallina* was investigated for the production of 1,2-epoxyoctane from 1-octene in a medium containing *n*-hexadecane [79]. An improved preparation of immobilized cells was achieved by using a mixed matrix of silicone polymer and alginate gel involving the aqueous medium. In the organic monophase, the cells immobilized by this method exhibited a higher accumulation of epoxide than free cells.

Immobilized cells of *Aureobasidium pullulans* in calcium alginate beads produced similar amounts of fructooligosaccharide from sucrose compared with co-existing free cells in a reactor [80], and immobilized cells of *Leuconostoc mesenteroides* produced dextran from sucrose [83]. In both cases, mass transfer within the immobilizing beads would be the rate-limiting step.

Gaseous fuels such as methane and hydrogen can be produced from waste products by microorganisms. Yang and Guo [87–88] reported methane production from whey-permeate by immobilized fermentative bacteria and methanogenic bacteria.

Table 7. Production of other useful chemicals by immobilized microbial cells

Product	Substrate	Support material	Microorganism	Ref.
1,2-Epoxyoctane	1-Octene	Ca-alginate & Silicon polymer	*Nocardia corallina*	[79]
Fructo -oligosaccharides	Sucrose	Ca-alginate	*Aureobasidium pullulans*	[80]
L-Phenylacetyl carbinol	Benzaldehyde	Ca-alginate	*Saccharomyces cerevisiae*	[81, 82]
Dextran	Sucrose	Celite, Alginate	*Leuconostoc mesenteroides*	[83]
Porphyrin		Agar, Carrageenan	*Rhodobacter sphaeroides*	[84]
NADH	NAD$^+$		*Arthrobacter* sp.	[85]
Invert sugar	Sucrose	Polyacrylamide	*Saccharomyces cerevisiae*	[86]

Various kinds of processes and experimental techniques have been investigated for other applications using immobilized microorganisms. For example, recent publications can be found on the system of continuously degrading limonen using entrapped *Rhodococcus fascians* for removing bitterness from citrus juice [89], immobilization of microsomes for glucuronides formation from drugs to clarify metabolic routes for the elimination of drugs [90], and glycosylation of ergot alkaloids [91].

2.2 Immobilized Plant Cells

Plant cells can produce various kinds of industrially important compounds such as alkaloids, glycosteroids, terpenoids, phenolics, enzymes, insecticides, pigments, and vitamins, which have complicated structures and cannot be produced by microorganisms. The immobilization of plant cells also has the same advantages as those of immobilized microorganisms. Since most of useful compounds produced by cultured plant cells are secondary metabolites, the effects of growth-limiting factors such as oxygen, nitrogen, phosphates, and sugars on the production rate are expected to be similar to those of the microbial production of antibiotics or enzymes. However, under the immobilized condition, plant cells are liable to form aggregates, which introduce a different ability, due to the cooperation between aggregate-constituent cells. Moreover, those factors, such as moderate stresses, e.g. shear stress and light irradiation, and chemical properties and/or the physical structure of immobilizing materials, are apt to change the production rate, growth rate, product secretion ratio, product yield, and metabolic pathway.

Recent studies on immobilized plant cells are summarized in Table 8. In the case of immobilization of *Catharanthus roseus* for the production of indole alkaloids, Facchini and DiCosmo [92–93] reported reduced productivity compared to that of free cells for ajmalicine production using glass fibers as an immobilizing

Table 8. Application of immobilized plant cells

Plant cell	Support material	Product	Ref.
Catharanthus roseus	Glass fibres	Catharanthine, Ajmalicine	[92, 93]
Catharanthus roseus	Non-woven polyester fiber	Indole alkaroids (Serpentine etc.)	[94, 95, 96]
Thalictrum rugosum	Glass fibres	Protoberberine alkaroids	[97]
Lithospermum erythrorhizon	Polyurethane	Shikonin	[98]
Lithospermum erythrorhizon	Ca-alginate	Shikonin	[99]
Coffea arabica	Polyurethane	Coffeins	[100]
Mucuna pruriens	Alginate	Catechols	[101]
Lavandula vera	Ca-alginate	Blue pigments	[102]
Solanum aviculare	Pectate	Verbenone etc.	[103]
Nicotiana tabacum	PLL-encapsulated Ca-alginate	Phenolics	[104]

material, while Archambault et al. [94–95] demonstrated that the release of serpentine increased by a factor of 10.

Production of shikonin, which was the first industrially produced compound using plant cultured cells, was investigated by Park et al. [98] with *Lithospermum erythrorhizon* cells immobilized on polyurethane foam matrices. They observed an enhanced production of shikonin. Kim and Chang [99] also studied shikonin production by calcium alginate entrapped cells and the increase in production by a factor of 2.5 by immobilization was demonstrated. And furthermore, with in situ extraction by *n*-hexadecane, the productivity increased to 7.4 times that of production in suspension cultures.

Most of the important products of plant cultured cells are stored in the cells and not spontaneously released. The permeabilization of products through cell membranes is one of the important problems for immobilized plant cell systems.

2.3 Immobilized Animal Cells

Large scale insect-cell culture systems have received considerable attention in the last few years. Insect cells such as those of silkworms, armyworms, loopers, a fruit flies, and mosquitoes are expected to be used as a host to baculoviruses and a more efficient system for in vitro recombinant protein production than that of mammalian cells because of the simpler culture media, relatively easy cell-line maintenance, immortality, relatively low sensitivity to stresses, smaller genome size, higher gene expression level etc. Agathos et al. [105] have reviewed free and immobilized insect cell cultures.

Most applications of animal cell cultures have been, however, those of mammalian cell cultures. Mammalian cells are used to produce pharmaceutically important human proteins such as interferons, lymphokines, tPA, urokinase, erythropoietin, hepatitis vaccines, and monoclonal antibodies.

Mammalian cells have a relatively high sensitivity to shear stress and are anchorage-dependent in many cases. Immobilization of mammalian cells has been studied to improve the efficiency and stability of their growth and the production of useful substances. Many review papers have been published [106–107]. Recent studies are shown in Table 9.

Various kinds of supporting materials are used for immobilization of mammalian cells. A considerable number of studies are still using calcium alginate as a supporting material [108, 110, 122]. However, other materials such as collagen [113], non-woven fabrics [120], chitosan-CMC [115], glass fibers [116–117], polyvinyl foamy resin [121] and polyurethane foam [118–119] or a composite of alginate and other materials such as chitosan-stabilized alginate [109], and alginate containing urethane foam [114] have been adopted or newly proposed. In order to select the most suitable immobilizing material for each cell line and product in relation to physical and chemical interactions between surfaces of materials and cells, more study is needed.

In immobilized hybridoma cultures, oxygen mass transfer is often a rate-limiting factor for antibody production. Wohlpart et al. [108] studied oxygen consumption

Table 9. Application of immobilized animal cells

Animal cell	Support material	Product	Ref.
Hybridoma 63D3	Ca-alginate	Monoclonal antibody	[108]
Hybridoma HP-6	Chitosan-stabilized Ca-alginate	Monoclonal antibody	[109]
Hybridoma S3H5/g2bA2	Alginate	Monoclonal antibody	[110]
Hybridoma C1E3	Porous spheres	Monoclinal antibody	[111]
Hybridoma HB124	Microporous hollow fiber	Monoclonal IgG	[112]
Hybridoma, CHO	Collagen microsphere	IgG	[113]
Hybridoma 16-3F	Alginate & urethane foam	Monoclonal antibody	[114]
Hybridoma HB-8852	Chitosan-CMC	Monoclonal antibody	[115]
Chinese hamster ovary	Glass fibers	γ-Interferon	[116, 117]
Monkey kidney, CHO	Polyurethane foam (PUF)	–	[118, 119]
Human embryonic lung diploid fibroblast	Nonwoven fabrics	tPA	[120]
Mouse myeloma MPC-11	Polyvinyl formal resin	–	[121]
Mouse neuroblastoma	Ca-alginate	–	[122]

by hybridoma cells immobilized in calcium alginate beads. The specific oxygen consumption rate of immobilized cells was higher than that of the free cells, while the growth rate of immobilized cells was reduced and the specific antibody production rate was comparable to that of free cells. Overgaard et al. [109] calculated the oxygen mass-transfer limitations within the immobilizing gel beads. Racher et al. [111] also concluded that the oxygen diffusion resistance into supporting spheres was a regulatory factor in the destabilization of antibody production. Chiou et al. [116–117] investigated a fiber-bed bioreactor for anchorage-dependent animal cell cultures. They studied strategies to design and operate this kind of reactor based on oxygen mass-transfer, and estimated a scale up potential of this reactor. Kamihira et al. [114] reported that increased antibody production was observed by changing the aeration gas to oxygen instead of air.

On the other hand, the improved stability of antibody production in a serum-free media by using immobilized cells was reported [110]. The effects of immobilization on the cell growth and the metabolism are expected to be examined.

2.4 Characteristics of Immobilized Whole Cells

From an engineering point of view, it is rather important to analyze rate-limiting factors of the desired products in the immobilized cell reaction system. In the systems using immobilized aerobic living cells, i.e. aerobic bacteria cells, plant cells, and animal cells, the process performance is usually determined by the oxygen mass-transfer limitation. In the cases of immobilized anaerobic bacteria, mass-transfer of a substrate for the growth and/or the reaction is often rate-limiting. In particular, intraparticle diffusion effects on the reactivity of immobilized whole cell systems are important. Many investigations on these mass-transfer phenomena in immobilized systems have been carried out hitherto as described above. However, few studies of these phenomena were thought to be sufficiently quantitative to evaluate the performance of the immobilized-cell processes.

For the investigation of intraparticle diffusion of oxygen, an oxygen microsensor is a convenient tool to measure directly the oxygen profile in immobilizing gel matrices [123-124].

On the other hand, the effects of mass transfer limitation in immobilizing particles are often estimated by combining diffusion coefficients with the reaction kinetics of cell growth and the product formation. Effective diffusivity in the immobilizing particles is measured using various methods [125-127]. The effect of cell concentration in immobilizing particles with respect to the effective diffusivity is important to analyze the phenomena relating immobilized growing cells. Many researchers have studied this problem [128-130, 131-136]. For the estimation of the intraparticle diffusional effects on immobilized growing cells, this problem must be studied more precisely [137-140].

In the near future, genetically engineered microorganisms will be used more often in various biological processes. A higher plasmid stability than free cells is often observed in immobilized recombinant microorganisms. Kumar and Schügerl have reviewed this problem [141]. Several groups of researchers have been studying the effects of immobilization procedures and culture conditions on plasmid stability [142-144].

3 Intraparticle Diffusion Effect on Immobilized Microbial Cells

A great number of biological processes have been investigated in order to utilize fully immobilized biocatalysts. However, some important problems caused by immobilization still remain to be solved, for example, the increased mass-transfer resistance [128-130], the changing of reaction properties, the treatment of deactivated biocatalysts, and the additional cost of immobilization. Particularly, in the case of immobilized growing whole cells, their growth rate in the immobilizing particles decreases with the growth of the cells because the diffusion limitation increases with cell growth [145]. Thus, it is very difficult to estimate the overall reaction rate of immobilized growing cells since the cell population and profile change with time in a complicated manner.

The purpose of this section is to elucidate microbial growth and reaction characteristics within the biocatalyst-immobilizing particles using the ethanol production system by *Zymomonas mobilis*, to clarify the effect of mass-transfer resistance on the cell growth, and the overall reactivity of the particles. Furthermore, the effect of co-immobilization of porous solid particles with microorganisms is discussed.

3.1 Materials and Methods

3.1.1 Microorganism and Medium

Zymomonas mobilis NRRL B14023 was used for our study. Culture conditions and the media components have been presented in a previous paper [21].

3.1.2 Cell Immobilization

The harvested microbial cells were suspended in a medium without glucose and mixed with $40 \, g \, l^{-1}$ sodium alginate solution of the same volume. The mixture obtained was dripped through a needle into the $15 \, g \, l^{-1}$ $CaCl_2 \cdot 2 \, H_2O$ solution gently stirred at $5 \, °C$.

In the case of immobilization with porous particles, the particles were mixed with an alginate solution in advance. In these particular experiments, two kinds of porous particles, silica gel particles with an average diameter of 10 µm and an average pore size of 5 nm (Wako Pure Chemicals, Osaka), and hydroxyapatite particles with an average diameter of 40 µm and an average pore size of 10–20 nm (Koken Co., Tokyo) were used.

The volume fraction of porous particles in the immobilizing gel particle was approximately 10–15%.

3.1.3 Measurement of Inherent Growth Rate

The growth rate of free cells was measured during the initial exponential phase of growth. The harvested cells were inoculated into the medium containing $0–240 \, g \, l^{-1}$ glucose, $0–104 \, g \, l^{-1}$ ethanol, $2.5 \, g \, l^{-1}$ yeast extract (Difco), $1 \, g \, l^{-1}$ KH_2PO_4, $1 \, g \, l^{-1}$ $(NH_4)_2SO_4$, $0.5 \, g \, l^{-1}$ $MgSO_4 \cdot 7 \, H_2O$, and $2.5 \, g \, l^{-1}$ $CaCl_2 \cdot 2 \, H_2O$. The time-course of the cell growth in the media of various glucose and ethanol concentrations in a 500-ml Erlenmeyer flask agitated with a magnetic stirrer was measured. From the cell growth curve, the inherent specific growth rates at the corresponding glucose and ethanol conentrations were calculated.

3.1.4 Measurement of the Growth Rate of Immobilized Cells

The growth rate of cells immobilized in particles was measured by a continuous stirred reactor maintained at $30 \, °C$. About 100 of the immobilized cell beads were introduced into 100–200 ml of growth medium in a 500-ml Erlenmeyer flask or a 200-ml cylindrical vessel agitated with a magnetic stirrer. The fresh medium was continuosly fed to the flask and the same volume of the medium overflowed. The pH of the medium in the vessel was kept constant at 5.0 using NaOH.

The cell distribution profile in the gel particle was measured by a scanning electron microscope [145]. In this measurement, the relative cell concentration at a certain position in the particle was determined by calculating the ratio of the area occupied by cell colonies to the total area from a microphotograph of a section including the position. Here, the cell density inside the colonies in the particles was assumed constant. The absolute value of the cell concentration profile was then calculated using the volume-averaged cell concentration in the particles obtained by dissolving the alginate gel particles with 10% sodium polyphosphate aqueous solution.

3.1.5 Analysis

Glucose concentrations were determined by the glucose oxidase method using a diagnostic kit, Glucose B-Test WAKO (Wako Pure Chemicals, Osaka). The ethanol concentration was determined by gas chromatography using Shimadzu 3BF with an FID detector. The cell concentration in the liquid phase was determined by a measurement of the optical density at 660 nm.

3.2 Cell Growth Rate in the Cell-Immobilizing Particles

3.2.1 Inherent Growth Rate

The specific growth rate of free *Z. mobilis* was measured at various glucose concentrations up to about 240 g l^{-1} [21]. It was assumed to be identical to the inherent growth rate of cells immobilized in the gel. The specific growth rate increased with glucose concentration up to 100 g l^{-1}, but the growth rate decreased linearly at higher concentrations due to substrate inhibition. Using the Monod-type kinetics for substrate limitation and linear substrate inhibition at $S_0 \geqq 100$ g l^{-1}, the specific growth rate was formulated by the following equation:

$$\mu = I_s(S) \, \mu_{max} S/(K_s + S) \tag{1}$$

where

$$I_s(S) = \begin{cases} 1; & S \leqq S_i \\ (S_{max} - S)/(S_{max} - S_i); & S_i \leqq S \leqq S_{max} \\ 0; & S_{max} \leqq S \end{cases} \tag{2}$$

Here, the maximum specific growth rate, μ_{max}, was 0.443 h^{-1}; the substrate limitation (saturation) constant, K_s, was 4.8 g l^{-1}; the maximum glucose concentration of cell growth, S_{max}, was 238 g l^{-1}; and the threshold glucose concentration for substrate inhibition, S_i, was 100 g l^{-1}.

The experimental results of the effect of ethanol concentration on the specific growth rate were obtained in the same way. The product inhibition could be expressed by two lines with different slopes. The following equation was obtained:

$$\mu = \mu_0 I_p(P) \tag{3}$$

where

$$I_p(P) = \begin{cases} 1 - P/K_p; & P \leqq P_i \\ (1 - P_i/K_p) \, (P_{max} - P)/(P_{max} - P_i); & P_i \leqq P \leqq P_{max} \\ 0; & P_{max} \leqq P \end{cases} \tag{4}$$

Here, the product (ethanol) inhibition constant, K_p, was 131 g l^{-1}; the maximum ethanol concentration for cell growth, P_{max}, was 91.7 g l^{-1}; and the transition ethanol concentration for product inhibition, P_i, was 60 g l^{-1}.

Rogers et al. [146] and Lee et al. [147–148] have reported equations of the growth rate of *Z. mobilis* ZM4 taking into account the substrate and product inhibition. Their equations are about the same as Eqs. (1)–(4) and their value of μ_{max} is 0.5 h^{-1}, which is considerably close to ours. Their results, however, included the appearance of the threshold ethanol concentration, that is, below this value, no product inhibition against growth rate was observed. They also showed that the substrate inhibition against cell-growth is milder than that in our case at a glucose concentration greater than 200 g l^{-1}. These differences may be caused by the rapid change of ethanol and glucose concentration in our study.

3.2.2 Cell-Growth Rate in the Cell-Immobilizing Particles

The transitional growth behavior of *Z. mobilis* inside the Ca-alginate gel particle was studied using a scanning electron microscope. An example is shown in Fig. 1. In this example, the average radius of gel particles was 1.35 mm after cultivation and the initial cell concentration was 0.43 g dry cell per l. The glucose concentration in the bulk solution was kept at 50 g l^{-1} throughout the experiment. During the whole time course of the cultivation, cell growth was observed everywhere inside the particles. The cell concentration decreased towards the center of the particles. Close to the surface, most of the microphotograph was occupied by cell colonies.

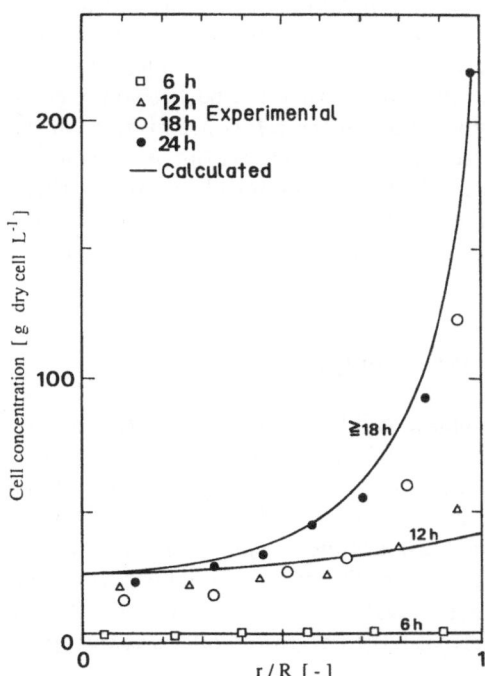

Fig. 1. Cell growth in the immobilized state in gel particles: Calculated results and experimental data :R = 1.35 mm, X_0 = 0.43 g dry cell per l, s_0 = 50 g l^{-1}

In this region, the cell concentration seemed to be very high, considering the physical growth limit. Thus, it was assumed that the maximum cell concentration there was 220 g dry cell per l, which was obtained from the experimental data of the close-packed volume of wet cells and their dry-cell weight.

3.3 Simulation of Cell-Growth Inside the Particles

The distribution profile of cell concentration along the radius of the cell-immobilizing gel particle and its variation were calculated numerically using the fundamental equations described in the Appendix. The assumptions underlying the calculations are as follows: First, the growth and reaction rates are the same as those of free cells. Here, the specific production rate follows the equation described in a previous paper [130]. The substrate inhibition of ethanol production was formulated by an exponential function for the convenience of calculation. Mass-transfer resistance outside the particles was considered negligible. Particles were considered spherical. The distribution coefficients of glucose and ethanol between liquid and gel phase are approximated to unity. Diffusion coefficients for glucose and ethanol follow Eq. (1) of reference [130]. And finally, the yield of ethanol with respect to the glucose consumption, $Y_{p/s}$, is considered to be a constant value of 0.48 g EtOH per g glucose at any position inside the particle.

In Fig. 1, an example of the calculated results of the distribution profile of the cell concentration is compared with the experimental results. The experimental data agreed approximately with the simulated results. However, the growth level in the particles 18 hours after the start of the cultivation seems to be a little lower than the simulated values. This phenomenon might be attributed to the effect of the pH change [149] inside the particles caused by carbon dioxide formation. Another cause for the time lag in the cell growth inside the particles may come from an unaccounted factor on the dependency of the cell concentration on the effective diffusivity. That is, Eq. (A9) (see the Appendix) was obtained from the experiments [130] using *Z. mobilis* cells distributed uniformly in the particle, although the cells formed colonies when they grew in the particles. Thus, the diffusivity inside particles containing cells that form colonies needs to be investigated further.

3.4 Overall Production Rate of the Immobilized Cells

The overall reaction rate of the immobilized cell biocatalyst was determined by the cell distribution profile in the particles. After a certain period of cultivation, the overall rate reaches a steady state. However, an interesting feature of the overall production rate is expected to be observed in the transition state of the cell-growth in the particle. The calculated result of the overall ethanol production rate for cell-containing particles with 2.0 mm radius and an initial concentration of $1.0 \, \text{g} \, \text{l}^{-1}$ is shown in Fig. 2. In each case for the various bulk glucose concentrations, the overall rate first increased, then decreased after a critical peak toward a constant value representing the steady state. Figure 3 shows the results

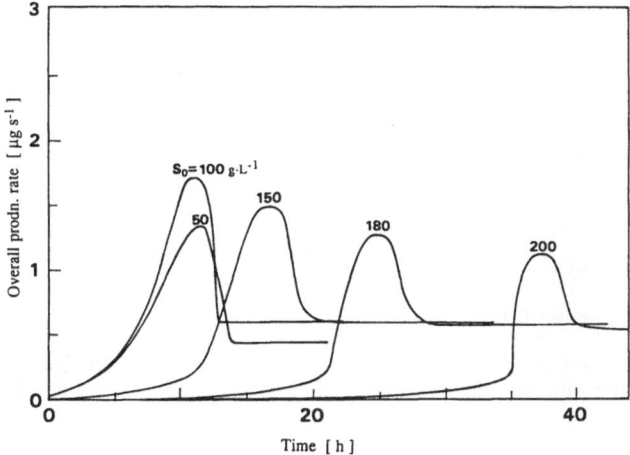

Fig. 2. Calculated results of variation of the overall production rate affected by the cell growth in the cell-immobilizing particles: $R = 2.0$ mm, $X_0 = 1.0$ g l^{-1}

of preliminary experiments using cell containing particles of 1.3–1.8 mm radii at initial cell concentrations of 0.41–4.25 g l^{-1}. Since various conditions of the particle preparation were employed in these experiments, time is normalized by the time when the production rate peaked. A tendency similar to that showed in Fig. 2 was observed, but the peak was not as sharp as expected from the calculations.

3.5 Effect of Co-Immobilization of Porous Particles with Cells

The start-up dynamics of the overall production rate of whole-cell immobilizing particles was examined in the previous section. The overall production rate first increased, then decreased after a critical peak, this phenomenon means that, if the

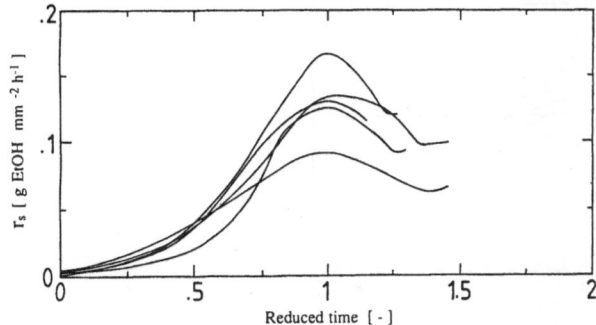

Fig. 3. Time courses of the overall production rate per unit surface area of cell-immobilizing particles: Experimental results under various conditions: $S_0 = 50$ g l^{-1}, $R = 1.3–1.8$ mm, $X_0 = 0.41–4.25$ g l^{-1}.

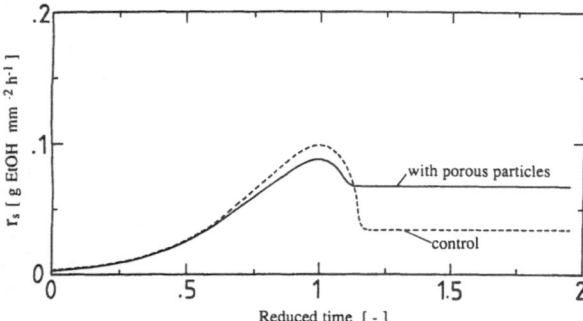

Fig. 4. Time courses of the overall production rate per unit surface area of cell-immobilizing particles: Calculated results for co-immobilization: $S_0 = 50\,g\,l^{-1}$, $R = 2.0\,mm$, $X_0 = 1.0\,g\,l^{-1}$

cells immobilized in the particles proliferate excessively, the cell growth would reduce the overall production rate because of a diffusional limitation caused by a volume fraction occupied by cells in the particles.

A co-immobilization of porous particles which have a far smaller pore size than the diameter of immobilized cells would reduce the above mass-transfer limitation. These co-immobilized porous particles could affect the overall production rate by decreasing the maximum cell concentration in the particles and by securing a diffusional route. An example of calculated results of a production rate by co-immobilized cells with porous particles is shown in Fig. 4. In this calculation, the volume fraction of porous particle in the cell-immobilizing particle was 0.2 and the ratio of the effective diffusivity in the porous particle to the diffusion coefficient in water was assumed to be 0.2, and an initial cell concentration was $1\,g\,l^{-1}$. In the case of co-immobilization with porous particles, the steady-state production rate was at a considerably higher level than in the case of immobilization

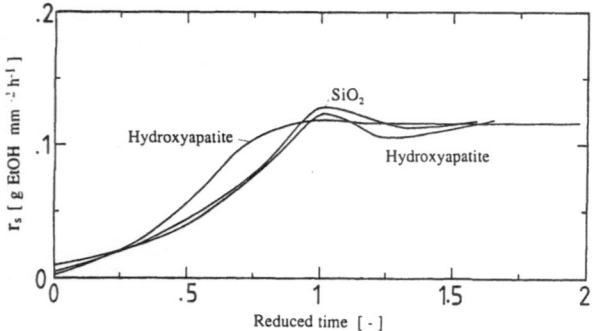

Fig. 5. Time courses of the overall production rate per unit surface area of cell-immobilizing particles. Cells were co-immobilized with porous particles. Experimental results for various porous particles: $S_0 = 50\,g\,l^{-1}$, $R = 1.2 - 2.0\,mm$, $X_0 = 0.6 - 1.4\,g\,l^{-1}$

without porous particles. Experimental results of co-immobilizing cells with porous particles are shown in Fig. 5. The overall production rate has an apparent tendency to stabilize at a higher level.

4 Future Prospects

Immobilized whole cell systems have many advantages over free cell systems as mentioned above. In particular, physiological effects caused by immobilization on the metabolic systems of living cells increase the possibilities of enhancing and/or stabilizing the production of useful chemicals. The effects of physical and/or chemical microenvironments around the immobilized cells on the cellular metabolic systems must be investigated more precisely. Studies of mutual interactions and symbiotic relations between different types of immobilized cells might offer some interesting applications of immobilized cell systems.

In the meantime, new biocatalytic activities of various kinds of cells have been developed. And furthermore, biocatalysts will be produced by new biotechnology including gene manipulation techniques, cell-fusion techniques, and protein engineering. Immobilization of these promising biocatalytic cells would also exploit the new world in the field of bioprocess engineering.

5 Appendix

The calculation of the cell growth inside the cell-immobilizing gel particles was carried out as described in the following paragraphs.

First, glucose and ethanol concentrations at position r and time t were calculated using the following equations:

(1) Mass balance of the substrate:

$$\frac{1}{r^2} \frac{\partial}{\partial r} \left(D_{es} r^2 \frac{\partial S}{\partial r} \right) - \frac{X}{Y_{p/s}} r_{pm} \frac{S}{K'_s + S} I'_s(S) I'_p(P) = 0 \tag{A1}$$

where

$$I'_s(S) = \begin{cases} 1 \, ; & S \leq S_i \\ \exp\left[-(S - S_i)/K_i\right] \, ; & S_i \leq S \end{cases} \tag{A2}$$

$$I'_p(P) = \begin{cases} 1 - P/K'_p \, ; & P \leq P'_i \\ (1 - P'_i/K'_p)(P'_{max} - P)/(P'_{max} - P'_i) \, ; & P'_i \leq P \leq P'_{max} \\ 0 \, ; & P'_{max} \leq P \end{cases} \tag{A3}$$

$r_{pm} = 4.19$ g EtOH per g dry cell per h, $\quad K'_s = 0.25 \, \text{g} \, 1^{-1}$,

$S_i = 100 \, \text{g} \, 1^{-1}$, $\quad K_i = 32.5 \, \text{g} \, 1^{-1}$, $\quad P'_i = 10 \, \text{g} \, 1^{-1}$,

$K'_p = 35.2 \, \text{g} \, 1^{-1}$, $\quad Y_{p/s} = 0.48$ g EtOH per g glucose, and

$P'_{max} = 128.8 \, \text{g} \, 1^{-1}$.

Boundary conditions:

$$S = S_0 \quad \text{at} \quad r = R \tag{A4}$$

$$\frac{\partial S}{\partial r} = 0 \quad \text{at} \quad r = 0 \tag{A5}$$

(2) Mass balance of the product:

$$\frac{1}{r^2} \frac{\partial}{\partial r} \left(D_{ep} r^2 \frac{\partial P}{\partial r} \right) + X r_{pm} \frac{S}{K'_s + S} I'_s(S) I'_p(P) = 0 \tag{A6}$$

Boundary conditions:

$$P = 0 \quad \text{at} \quad r = R \tag{A7}$$

$$\frac{\partial P}{\partial r} = 0 \quad \text{at} \quad r = 0 \tag{A8}$$

(3) Diffusivity

$$D_{es}/D_{0s} = D_{ep}/D_{0p} = k_1 (1 - k_2 X)^2 \tag{A9}$$

where

$$k_1 = 0.68, \quad k_2 = 0.004 \, l \, g^{-1}$$
$$D_{0s} = 3.1 \times 10^{-6} \, m^2 \, h^{-1}, \quad D_{0p} = 4.1 \times 10^{-6} \, m^2 \, h^{-1}$$

The calculations were carried out using the shooting method, that is, iterations by changing the assumed value of $\partial S/\partial r$ at $r = R$ until the conditions at $r = 0$ were satisfied. The integration was carried out using the Runge-Kutta method.

Secondly, the growth rates at different positions along the radial coordinate inside the particles were calculated using the Eqs. (1)–(4). The cell concentration at time $t + dt$ was then obtained using the growth rate at time t.

Finally, the cell concentration in the particle after a certain time was obtained by repeating the first and second procedures described here.

6 References

1. Kennedy JF, Cabral JMS (1983) Appl Biochem Bioeng 4: 189
2. Corcoran E (1986) Top Enzyme Ferment Biotechnol 10: 12
3. Chibata I, Tosa T, Sato T (1986) J Mol Catal 37: 1
4. Scott CD (1987) Enzyme Microb Technol 9: 66
5. Furusaki S (1988) J Chem Eng Japan 21: 219
6. Klein J, Kressdorf B (1989) Angew Makromol Chem 166–167: 293

7. Tanaka A, Nakajima H (1990) Adv Biochem Eng/Biotechnol, 42: 97
8. Woodward J (ed) (1985) Immobilized cells and enzymes. IRL Press, Oxford (Practical Approach Series)
9. Moo-Young M (ed) (1988) Bioreactor immobilized enzymes and cells. Elsevier Applied Science, London
10. Mosbach K (ed) (1987–8) Methods in Enzymol 135–137 (Immobilized Enzymes and Cells Pt B, C, D)
11. Bravo P, Gonzalez G (1991) J Chem Technol Biotechnol 52: 127
12. Gil GH, Jones WJ, Tornabene TG (1991) Enzyme Microb Technol 13: 390
13. Ramakrishna SV, Prema P, Sai PST (1991) Bioprocess Eng 6: 117
14. Chen C, Dale MC, Okos MR (1990) Biotechnol Bioeng 36: 975
15. Joung JJ, Royer GP (1990) Ann NY Acad Sci 589: 271
16. Amin G, Doelle HW (1990) Enzyme Microbial Technol 12: 443
17. Galazzo JL, Bailey JE (1990) Biotechnol Bioeng 36: 417
18. Hilge-Rotmann B, Rehm HJ (1990) Appl Microbiol Biotechnol 33: 54
19. Dale MC, Chen C, Okos MR (1990) Biotechnol Bioeng 36: 983
20. Chen C, Dale MC, Okos MR (1990) Biotechnol Bioeng 36: 993
21. Seki M, Furusaki S, Shigematsu K (1990) Ann NY Acad Sci 613: 290
22. Monbouquette HG, Sayles GD, Ollis DF (1990) Biotechnol Bioeng 35: 609
23. Hannoun BJM, Stephanopoulos G (1990) Biotechnol Prog 6: 341
24. Hannoun BJM, Stephanopoulos G (1990) Biotechnol Prog 6: 349
25. Bringi V, Dale BE (1990) Biotechnol Prog 6: 205
26. Hilge-Rotmann B, Rehm HJ (1991) Appl Microbiol Biotechnol 34: 502
27. Hamada T, Sugishita M, Motai H (1990) Appl Microbiol Biotechnol 33: 624
28. Groot WJ, den Reyer MCH, Baart de la Faille T, van der Lans RGJM, Luyben KChAM (1991) Chem J (Lausanne) 46: B1
29. Park CH, Okos MR, Wankat PC (1990) Biotechnol Bioeng 36: 207
30. Friedl A, Qureshi N, Maddox IS (1991) Biotechnol Bioeng 38: 518
31. Hecker D, Bisping B, Rehm HJ (1990) Appl Microbiol Biotechnol 32: 627
32. Bisping B, Baumann U, Rehm HJ (1990) Appl Microbiol Biotechnol 32: 380
33. Hoetmann U, Bisping B, Rehm HJ (1991) Appl Microbiol Biotechnol 35: 258
34. Steinmeyer DE, Shuler ML (1990) Biotechnol Prog 6: 286
35. Melzoch K, Rychtera M, Markvichov NS, Pospichalova V, Basarova G, Manakov MN (1991) Appl Microbiol Bioeng 34: 469
36. Godia F, Adler HI, Scott CD, Davison BH (1990) Biotechnol Prog 6: 210
37. Krischke W, Schroeder M, Troesch W (1991) Appl Microbiol Biotechnol 34: 573
38. Roukas T, Kotzekidou P (1991) Enzyme Microb Technol 13: 33
39. Audet P, Paquin C, Lacroix C (1990) Appl Microbiol Biotechnol 32: 662
40. Yabannavar VM, Wang DIC (1991) Biotechnol Bioeng 37: 544
41. Hoshino K, Taniguchi M, Marumoto H, Shimizu K, Fujii M (1991) Agric Biol Chem 55: 479
42. Dinarieva T, Netrusov A (1991) Biotechnol Prog 7: 234
43. Crespo JPSG, Almeida JS, Moura MJ, Carrondo MJT (1990) Biotechnol Bioeng 36: 705
44. Kautola H, Rymowicz W, Linko YY, Linko P (1991) Appl Microbiol Biotechnol 35: 447
45. O'Brien DJ, Panzer CC, Eisele WP (1990) Biotechnol Prog 6: 237
46. Sun Y, Furusaki S (1991) J Ferment Bioeng 69: 102
47. Sun Y, Furusaki S (1991) J Ferment Bioeng 70: 196
48. Lacroix C, Paquin C, Arnaud JP (1990) Appl Microbiol Biotechnol 32: 403
49. Constantinides A, Mehta N (1991) Biotechnol Bioeng 37: 1010
50. Keshavarz T, Eglin R, Walker E, Bucke C, Holt G, Bull AT, Lilly MD (1990) Biotechnol Bioeng 36: 763
51. Chun GT, Agathos SN (1991) Biotechnol Bioeng 37: 256
52. Trueck HU, Chmiel H, Hammes WP, Troesch W (1990) Appl Microbiol Biotechnol 34: 1
53. Trueck HW, Chmiel H, Hammes WP, Troesch W (1990) Appl Microbiol Biotechnol 33: 139
54. Sachse H, Kude J, Kerns G, Berger R (1990) Acta Biotechnol 10: 523

55. Gallo Federici R, Federici F, Petruccioli M (1990) Biotechnol Lett 12: 661
56. Federici F, Petruccioli M, Miller MW (1990) Appl Microbiol Biotechnol 33: 407
57. Fiedurek J, Lobarzewski J (1990) Starch 42: 358
58. Kuek C (1991) Appl Microbiol Biotechnol 35: 466
59. Klingeberg M, Vorlop KD, Antranikian G (1990) Appl Microbiol Biotechnol 33: 494
60. Johri BN, Alurralde JD, Klein J (1990) Appl Microbiol Biotechnol 33: 367
61. Fortin C, Vuillemard JC (1990) Biotechnol Lett 12: 913
62. El-Aassar SA, El-Badry HM, Abdel-Fattah AF (1990) Appl Microbiol Biotechnol 33: 26
63. Kily T, Tiedtke A (1991) Appl Microbiol Biotechnol 35: 14
64. Cornwell KL, Tinland-Butez MF, Tardone PJ, Cabasso I, Hammel KE (1990) Enzyme
    Microb Technol 12: 916
65. Chen AHC, Dosoretz CG, Grethlein HE (1991) Enzyme Microb Technol 13: 404
66. Kirkpatrick N, Reid ID, Ziomek E, Paice MG (1990) Appl Microbiol Biotechnol 33: 105
67. Kawabata N, Nakagawa K (1991) J Ferment Bioeng 71: 19
68. Vlahov R, Pramatarova V, Spassov G, Suchodolskaya GV, Koshcheenko KA (1990)
    Appl Microbiol Biotechnol 33: 172
69. Hocknull MD, Lilly MD (1990) Appl Microbiol Biotechnol 33: 148
70. Abramov S, Aharonowitz Y, Harnik M, Lamed R, Freeman A (1990) Enzyme Microb
    Technol 12: 982
71. Macaskie LE, Dean ACR (1990) Applied Microbiol Bioeng 33: 81
72. Ferschl A, Loidl M, Ditzelmueller G, Hinteregger C, Streichsbier F (1991) Appl
    Microbiol Biotechnol 35: 544
73. Lvingston AG, Willacy A (1991) Appl Microbiol Biotechnol 35: 551
74. Balfanz J, Rehm HJ (1991) Appl Microbiol Biotechnol 35: 662
75. Beunink J, Rehm HJ (1990) Appl Microbiol Biotechnol 34: 108
76. Moersen A, Rehm HJ (1990) Appl Microbiol Biotechnol 33: 206
77. Wijffels RH, Schukking GC, Tramper J (1990) Appl Microbiol Biotechnol 34: 399
78. Hooijmans CM, Geraats SGM, van Neil EWJ, Robertson LA, Heijnen JJ, Luyben
    KChAM (1990) Biotechnol Bioeng 36: 931
79. Kawakami K, Tsuruda S, Miyagi K (1990) Biotechnol Prog 6: 357
80. Yun JW, Jung KH, Oh JW, Lee JH (1990) Appl Biochem Biotechnol 24–25: 299
81. Mahmoud WM, El-Sayed HMM, Coughlin RW (1990) Biotechnol Bioeng 36: 55
82. Mahmoud WM, El-Sayed AHMM, Coughlin RW (1990) Biotechnol Bioeng 36: 47
83. El-Sayed AHMM, Mahmoud WM, Coughlin RW (1990) Biotechnol Bioeng 36: 83
84. Ishii K, Hiraishi A, Arai T, Kitamura H (1990) J Ferment Bioeng 69: 26
85. Nath PK, Izumi Y, Yamada H (1990) Enzyme Microb Technol 12: 28
86. Ghose S (1990) Biotechnol Tech 4: 419
87. Yang ST, Guo M (1990) Biotechnol Bioeng 36: 427
88. Yang ST, Guo M (1991) Biotechnol Bioeng 37: 375
89. Manjon A, Iborra JL, Martinez-Madrid C (1991) Appl Microbiol Biotechnol 35: 176
90. Haumont M, Magdalou J, Ziegler JC, Bidault R, Siest JP, Siest G (1991) Appl Microbiol
    Biotechnol 35: 440
91. Kren V, Flieger M, Sajdl P (1990) Appl Microbiol Biotechnol 32: 645
92. Facchini PJ, DiCosmo F (1991) Appl Microbiol Biotechnol 35: 382
93. Facchini PJ, DiCosmo F (1991) Appl Microbiol Biotechnol 33: 36
94. Rho D, Bedard C, Archambault J (1990) Appl Microbiol Biotechnol 33: 59
95. Archambault J, Volesky B, Kurz WGW (1990) Biotechnol Bioeng 35: 702
96. Archambault J, Volesky B, Kurz WGW (1990) Biotechnol Bioeng 35: 660
97. Facchini PJ, DiCosmo F (1991) Biotechnol Bioeng 37: 397
98. Park YH, Seo WT, Liu JR (1990) J Ferment Bioeng 70: 317
99. Kim DJ, Chang HN (1990) Biotechnol Bioeng 36: 460
100. Furuya T, Koge K, Orihara Y (1990) Plant Cell Rep 9: 125
101. Pras N, Booi GE, Dijkstre D, Horn AS, Malingre TM (1990) Plant Cell, Tissue Organ
     Cult 21: 9
102. Nakajima H, Sonomoto K, Sato F, Yamada Y, Tanaka A (1990) Agric Biol Chem 54: 53
103. Vanek T, Macek T, Stransky K, Ubik K (1989) Biotechnol Tech 3: 411

104. Haigh JR, Linden JC (1989) Plant Cell Rep 8: 475
105. Agathos SN, Jeong YH, Venkat K (1990) Ann NY Acad Sci 589: 372
106. Looby D, Griffiths B (1990) Trends Biotechnol 8: 204
107. Goosen MFA (1988) Chem Eng Educ 22: 196
108. Wohlpart D, Gainer J, Kirwan D (1991) Biotechnol Bioeng 37: 1050
109. Overgaard S, Scharer JM, Moo-Young M, Bols NC (1991) Can J Chem Eng 69: 439
110. Lee GM, Palsson BO (1990) Biotechnol Bioeng 36: 1049
111. Racher AJ, Looby D, Griffiths JB (1990) J Biotechnol 15: 129
112. Heath C, Dilwith R, Belfort G (1990) J Biotechnol 15: 71
113. Ray NG, Tung AS, Hayman EG, Vournakis JN, Runstadler PW Jr (1990) Ann NY Acad Sci 589: 443
114. Kamihira M, Yoshida H, Iijima S, Kobayashi T (1990) J Ferment Bioeng 69: 311
115. Yoshioka T, Hirano R, Shioya T, Kako M (1990) Biotechnol Bioeng 35: 66
116. Murakami S, Chiou TW, Wang DIC (1991) Biotechnol Bioeng 37: 762
117. Chiou TW, Murakami S, Wang DIC (1991) Biotechnol Bioeng 37: 755
118. Matsushita T, Ketayama M, Kamihata K, Funatsu K (1990) Appl Microbiol Biotechnol 33: 287
119. Matsushita T, Hidaka H, Kamihata K, Kawakubo Y, Funatsu K (1991) Appl Microbiol Biotechnol 35: 159
120. Mitsuda S, Matsuda Y, Itagaki Y, Suzuki A, Kumazawa E, Higashio K, Kawanishi G (1990) J Ferment. Bioeng 70: 289
121. Yamaji H, Fukuda H (1991) Appl Microbiol Biotechnol 34(6): 730
122. Tamponnet C, Boisseau S, Lirsac PN, Barbotin JN, Poujeol C, Lievremont M, Simonneau M (1990) Appl Microbiol Biotechnol 33: 442
123. Hoojimans CM, Briasco CA, Huang J, Geraats BGM, Barbotin JN, Thomas D, Luyben KChAM (1990) Appl Microbiol Biotechnol 33: 611
124. Huang J, Hooijmans CM, Briasco CA, Geraats SGM, Luyben KChAM, Thomas D, Barbotin JN (1990) Appl Microbiol Biotechnol 33: 619
125. Westrin BA, Axelsson A (1991) Biotechnol Bioeng 38: 439
126. Westrin BA (1990) Appl Microbiol Biotechnol 34: 189
127. Westrin BA (1990) Biotechnol Tech 4: 409
128. Klein J, Schara P (1981) Appl Biochem Biotechnol 6: 91
129. Furusaki S, Seki M (1985) J Chem Eng Jpn 18: 389
130. Sakaki K, Nozawa T, Furusaki S (1988) Biotechnol Bioeng 31: 603
131. Sun Y, Furusaki S (1989) Biotechnol Bioeng 34: 55
132. Furui M, Yamashita K (1985) J Ferment Technol 63: 167
133. Pu HT, Yang RYK (1988) Biotechnol Bioeng 32: 891
134. Chresand TJ, Dale BE, Hanson SL, Gillies RJ (1988) Biotechnol Bioeng 32: 1029
135. Ho CS, Ju LK (1988) Biotechnol Bioeng 32: 313
136. Mignot L, Junter GA (1990) Appl Microbiol Biotechnol 32: 418
137. Dalili M, Chau PC (1987) Appl Microbiol Biotechnol 26: 500
138. Sayles GD, Ollis DF (1990) Biotechnol Prog 6: 153
139. Ruggeri B, Gianetto A, Sicardi S, Specchia V (1991) Chem Eng J 46: B21
140. Furusaki S (1989) Intradiffusion effect on reactivity of immobilized microorganisms. In: Fiechter A et al (eds) Bioproducts and Bioprocesses Springer, Berlin Heidelberg New York p 71
141. Kumar PKR, Schügerl K (1990) J Biotechnol 14(3–4): 255
142. Huang J, Dhulster P, Thomas D, Barbotin JN (1990) Enzyme Microb Technol 12(12): 933
143. Barbotin JN, Sayadi S, Nasri M, Berry F, Thomas D (1990) Ann NY Acad Sci 613: 41
144. Berry F, Sayadi S, Nasri M, Thomas D, Barbotin JN (1990) J Biotechnol 16: 199
145. Seki M, Furusaki S (1985) J Chem Eng Jpn 18: 461
146. Rogers PL, Lee KJ, Tribe DE (1979) Biotechnol Lett 1: 165
147. Lee KJ, Tribe DE, Rogers PL (1979) Biotechnol Lett 1: 421
148. Lee KJ, Rogers PL (1983) Chem Eng J 27: B31
149. Bajpai PK, Margaritis A (1986) Biotechnol Bioeng 28: 824

Modelling the Growth of Filamentous Fungi

J. Nielsen
Department of Biotechnology, Technical University of Denmark,
DK-2800 Lyngby, Denmark

Despite the considerable industrial importance of filamentous fungi there have been very few attempts to model the complex growth process of these microorganisms. With a new generation of high performance, computerized bioreactors and new analytical techniques it is possible to obtain the necessary experimental data for setting up reliable structured models describing the growth process of filamentous fungi. It is therefore interesting to review the mathematical models described previously in the literature and the experimental data on which these models are built. Only structured models are considered due to the complex metabolism of filamentous fungi and to the natural cellular structuring of the biomass, i.e. the biomass can be divided into different cell types.

    In order to set up good structured models it is strictly necessary to have a detailed knowledge of the mechanisms underlying the growth process. This involves both biochemical insight and understanding of the interactions between different macromolecules and cytological organelles.

Advances in Biochemical Engineering/
Biotechnology, Vol. 46
Managing Editor: A. Fiechter
© Springer-Verlag Berlin Heidelberg 1992

List of Symbols and Abbreviations

b   breakage function (hyphal elements formed per m^3 per h)
d   hypha diameter (m)
D   dilution rate in the bioreactor (h^{-1})
e_L column vector of dimension L with all elements being 1
k   rate constant (h^{-1})
K   saturation constant (kg per kg DW)
l_{hgu} hyphal growth unit length (m per tip)
m   mass of hyphal element (kg)
M   diagonal matrix containing the specific growth rates of the morphological
         forms (h^{-1})
n   average number of tips in a hyphal element
r   specific rate vector for intracellular reactions (h^{-1})
r_{tip} tip extension rate (kg DW per tip per h)
r_x biomass formation rate (kg DW $m^{-3} h^{-1}$)
r_V specific rate of increase in the wall area $(m^2 h^{-1})$
r_{VSC} rate of displacement of the VSC $(m h^{-1})$
R   pellet radius (m)
s   extracellular substrate concentration (kg m^{-3})
S   intracellular substrate concentration (kg per kg DW)
u   diagonal matrix containing the rate of metamorphosis reactions (h^{-1})
V_{hgu} hyphal growth unit volume (m^3 per tip)
w   water content in the cells (kg per kg biomass)
x   biomass concentration (kg DW m^{-3})
x_{hgu} hyphal growth unit mass (kg DW per tip)
X   intracellular concentration vector (kg per kg DW)
Y_{ij} stoichiometric coefficients (mole j (mole i)$^{-1}$)
z_{sk} distance between apex and position of the VSC
Z   fractional concentration vector of morphological forms (kg per kg DW)

Greek Symbols

α   stoichiometric coefficients for the substrate
δ   stoichiometric coefficients in the metamorphosis reactions
Δ   matrix containing the stoichiometric coefficients δ
Γ   matrix containing the stoichiometric coefficients in the intracellular reac-
         tions
ε   number of hyphal elements
μ   specific growth rate (h^{-1})
φ   branching frequency (tips formed per h)
ϱ   cell density (kg m^{-3})

1 Introduction

Filamentous fungi comprise an industrially very important sub-group of the heterogeneous collection of microorganisms called fungi, since they are used for the production of a wide variety of products ranging from primary metabolites (e.g. citric acid) to secondary metabolites (e.g. antibiotics) and further on to industrial enzymes (e.g. proteases and lipases). It is therefore of considerable importance to understand the mechanisms underyling the growth and product formation of filamentous fungi, since such knowledge may be used in the optimization of the microbial process. Structured mathematical modelling of microorganisms has proven to be an excellent tool for examination of microbial physiology, since the model may guide the researcher to set up experiments that reveal new aspects of the microbial behavior. Furthermore, when a good model has been obtained this may be combined with traditional engineering methods for optimization, scale-up and process control.

With the large variety of products produced by filamentous fungi it is impossible to construct a model that describes product formation in general. For all microorganisms product formation is, however, in one sense or the other coupled to the growth process and setting up a growth model is therefore a good starting point for any quantitative description of a microbial process. In this presentation only the growth process is described whereas product formation is not considered (see e.g. Nüesch et al. [1] for a description of β-lactam biosynthesis). Despite large variations in microbial physiology of filamentous fungi, these organisms share a large number of characteristic growth mechanisms. It should therefore be possible to set up a generally structured model, which describes most of the observations made during growth of these microorganisms.

Growth of microorganisms involves uptake of substrates present in the medium and further conversion of these to intracellular components, and the biochemistry of the microorganism is a determining factor for the overall growth process. The mechanisms underlying growth of filamentous fungi are very different from that of unicellular microorganisms, e.g. the growth occurs only at the tips of the multicellular element. This presentation therefore contains a short review of both the physiology of filamentous fungi and of the underlying growth mechanisms.

2 Modelling of Microbial Processes

A bioprocess involving filamentous fungi is normally carried out as a submerged cultivation, and a description of the overall process therefore involves modelling of both the microbial kinetics, i.e. the biotic phase, and the bioreactor performance, i.e. the abiotic phase. Modelling of bioreactor performance involves description of concentration gradients in the bioreactor, which are caused by the flow pattern for the gas and the liquid phase. Furthermore description of mass transfer phenomena is an important part of a bioreactor model. Especially for processes involving filamentous fungi, it is necessary to describe the mass transfer of oxygen from the gas to the liquid phase, since these processes are often operated with oxygen transfer limitations due to the high viscosity of the reaction mixture (for a review of oxygen transfer to viscous media see Schügerl [2]). Furthermore substrate diffusion may be the rate limiting process when filamentous fungi are grown in the form of pellets. In the present paper the focus is on the microbial kinetics, i.e. on a description of how the microorganisms respond to changes in the environment. Influences of bioreactor performance on the overall process are not considered since these influences are treated at length elsewhere [3, 4].

The biotic phase in a bioreactor is made up of a population of individual cells and it is therefore necessary to combine a kinetic model for the individual cells with a population model when the overall reaction kinetics is to be quantified (Fig. 1). Very often it is assumed that all the cells in the population are identical, i.e. they can be described by the same state vector, and a very simple population model can be applied. When the diversity of the individual cells is described in the population model it is normally referred to as a segregated model (similarly the simple population model mentioned above is referred to as non-segregated). In many cases the kinetics of the individual cells is stated as a simple unstructured model [5], e.g. the Monod model. These models may describe balanced growth conditions quite well, but normally they fail to describe growth at transient operating conditions, e.g. fed-batch operation. By including information of the cellular state one may obtain a model which is generally valid, i.e. it describes growth under both steady state and transient operating conditions, and it therefore has more predictive strength than a traditional unstructured model. Microbial models where information of the cellular state is included are normally referred to as structured models.

The state of a microbial cell is determined exclusively by its composition and it is therefore theoretically possible to set up a structured model that describes the cellular behavior completely. Considering the large number of intracellular components and reactions this is, however, in practice an impossible task, and in most cases a good description of the cellular behavior is obtained by including only a few central intracellular components and reactions. The number of

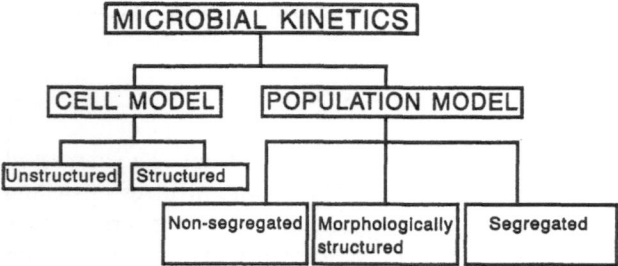

Fig. 1. Modelling of the microbial kinetics of the whole population involves description of the population and the individual microbial cells. The population model may either be non-segregated (all cells are identical), morphologically structured (a finite number of cell types considered), or segregated (a distribution of cells considered). The cellular kinetics may be described either by a structured or an unstructured model

intracellular components included in a structured model depends on the aim of the modelling work, and it may vary from 2–5 in simple structured models used for simulation of microbial processes to above 20 in single cell models used for examination of microbial behaviour at the cellular level [5]. For some microbial systems the exact mechanisms behind differentiation processes are not known, e.g. spore germination and sporulation, and it may therefore be valuable to describe the kinetics of each morphological form individually, followed by a description of conversion between the different morphological forms. This approach is referred to as morphological structuring [5], and it is a valuable extension of the traditional intracellularly structured model approach, especially for description of the growth of filamentous fungi. In a morphologically structured model we divide the cells into a finite number Q of morphological forms (or classes), and we assume that the intracellular composition is the same for all cells in each class. Besides being valuable for description of complex microbial systems these models also represent a link between the non-segregated and the segregated population models, i.e. when Q = 1 one has a non-segregated model and when Q → ∞ one has a segregated model (Fig. 1).

In the following the different aspects of modelling microbial kinetics are discussed briefly, whereas a more detailed description is given in references [5, 6, 7].

2.1 Cellular Kinetics

Cellular kinetics describes how the individual cells respond to changes in the environment, i.e. how the cells grow and form different products. With an intracellularly structured model, L key intracellular components are considered and the intracellular concentration of these components is given as the mass fraction of the total biomass, i.e. X_i is the concentration of the i'th component in kg i per kg biomass (as dry weight). The sum of all intracellular concentrations therefore equals 1. Changes in the concentrations of the intracellular components are described by specifying the stoichiometry for the intracellular reactions and the rates of the individual reactions. The growth process can in a natural way be divided into three types of reaction [8]: 1) Uptake of substrates, 2) conversion of intracellular substrates into cellular components and metabolic products, and 3) excretion of metabolic products. The stoichiometry for the general situation with N substrates, L intracellular components formed in J reactions, and M metabolic products is given elsewhere [5, 8]. The forward reaction rates of the J intracellular reactions are specified by the rate vector $\mathbf{r}$, and when the stoichiometric coefficients for the J intracellular reactions are collected in the matrix Γ the specific growth rate is calculated from Eq. (1).

$$\mu = \mathbf{e}_L^T \Gamma^T \mathbf{r} \quad (h^{-1}) \tag{1}$$

Equation (1) is based on the assumption that the substrate uptake and the product excretion reactions do not contribute to net formation of biomass [8]. For derivation of mass balances for the intracellular components see [5]. With an unstructured cellular model $L = 1$, and with only one reaction (1) reduces to $\mu = \gamma r$, where γ is a stoichiometric coefficient and r is a function of state variables in the abiotic phase only.

2.2 Population Models

In a segregated population model the distribution of microbial cells with a certain characteristic behavior (e.g. a certain content of a key intracellular component) is described [9, 10], whereas in a non-segregated population model all the microbial cells are assumed to be identical. The last type of model represents an average picture of the population, and it therefore gives less information. A serious drawback for the segregated models is the mathematical complexity, i.e. the mass balances are integro-differential equations, and one therefore tends to apply the much simpler non-segregated population model approach where the mass balances are trivial [5].

The morphologically structured models represent a compromise between the non-segregated and the segregated population models. Here a finite number of morphological forms are considered — each with a certain characteristic behavior. All the cells within each form are assumed to have the same intracellular composition. In the morphologically structured models the concentration of each morphological form is specified as the mass fraction of the total biomass, i.e. Z_q is the concentration of the q'th morphological form in kg q per kg biomass (dry weight).

The morphological form of a microbial cell is in principle determined by the intracellular composition, but it is often impossible to specify the exact composition that characterizes a given morphological form. The morphological form of a microbial cell is therefore often defined on the basis of simple macroscopic observations like cell size, cell shape, cellular position in a hypha etc. During microbial growth morphological forms are converted to each other, e.g. a spore is converted to an actively growing cell, and since each form is characterized completely by its intracellular composition, these conversions are determined exclusively by certain intracellular reactions. It is, however, impossible to specify all these intracellular reactions, and in a description of the conversion between different morphological forms we therefore introduce a set of K empirical reactions, called metamorphosis reactions [5]. Despite their empirical nature, the metamorphosis reactions may well have a biological meaning and they enable one to model complex systems where all the mechanisms are not yet known. The concept of morphological structuring may be combined with an intracellularly structured model as described by Nielsen and Villadsen [5].

With Q morphological forms and K metamorphosis reactions the stoichiometry can be summarized by Eq. (2).

$$\sum_{q=1}^{Q} \delta_{kq} Z = 0 \quad \text{forward reaction rate} \quad u_k \, (h^{-1}); \quad k = 1, ..., K$$

$$(2)$$

It is assumed that the spontaneous change of one morphological form into another does not involve any change in the total mass, and the sum of all stoichiometric coefficients in each reaction is therefore zero [5].

Often the specific growth rate of each morphological form (μ_q) is calculated from a simple unstructured model. However, if an intracellularly structured model is used to describe the specific growth rate of each morphological form, it is necessary to specify an intracellular state vector for each morphological form X_q, which contains the concentration of each intracellular component [5] (X_{iq} is the concentration of the i'th component in the q'th morphological form in kg i per kg q'th morphological form). Furthermore, it is necessary to specify the stoichiometry and the rates for the intracellular reactions in each morphological form. The specific growth rate for each morphological form may then be calculated from Eq. (1), and the specific growth rate of the total biomass is given by Eq. (3).

$$\mu = \sum_{q=1}^{Q} \mu_q Z_q \tag{3}$$

The formation rate of each morphological form is determined both by the metamorphosis reactions and by the growth associated reactions for each form, and it can be shown [5] that a mass balance for the morphological forms in a bioreactor with no inlet flow of microbial cells is given by Eq. (4), where M is a diagonal matrix $\{\mu_q\}$ and Δ contains the stoichiometric coefficients δ_{kq}.

$$\frac{dZ}{dt} = \Delta^T \mathbf{u} \mathbf{e}_K + MZ - \mu Z \tag{4}$$

3 Physiology of Filamentous Fungi

To model microbial growth one must have some knowledge of the microbial physiology, especially the microbial behavior under different operating conditions in bioreactors. One of the best experimental tools for this task is the steady state chemostat. This is illustrated by the extensive literature for the yeast *Saccharomyces cerevisae*, where steady state measurements of the concentrations of biomass, substrates, and metabolic products (both in the liquid and gas phase) has revealed many of the basic mechanisms behind growth and product formation. If the measurements of extracellular variables are combined with measurements of intracellular macromolecules much information on the basic cellular mechanisms may be obtained. This was illustrated by Måløe and Kjeldgaard [11] in studies of *Escherichia coli*, and their concept has been applied by many research groups in the field of bacterial growth (see a recent book by Cooper [12]).

Unfortunately the same approach has only rarely been applied to filamentous fungi, and this is a major reason for our present lack of information concerning the growth physiology of these microorganisms. In this section, some of the available experimental data giving information of the growth physiology are reviewed, both with respect to catabolism (energy forming reactions) and anabolism (biosynthetic reactions). A short discussion of available data for the macromolecular composition at different operating conditions is also given.

There are lots of good reasons for the paucity of steady state growth data for filamentous fungi, and it is often argued that it is impossible to obtain a true steady state due to the mycelial structure of these microorganisms. Basically, the problems are related to the non-Newtonian behavior of filamentous cultures, to wall growth, and to the extremely variable growth morphology. Many of the problems caused by the viscous nature of the reaction mixture can be eliminated by operating the chemostat at constant reactor mass rather than at constant volume. Furthermore, problems with wall growth can be minimized by reducing the head space volume to a minimum and by using a bioreactor with a low aspect ratio. With a modern high performance bioreactor fulfilling these demands we can obtain in our laboratory true steady states for a chemostat containing *Penicillium chrysogenum* (or at least we can claim that measurements of a large number of variables indicate that a true steady state is obtained).

3.1 Catabolism

Filamentous fungi may metabolize a wide variety of carbon sources, including many different carbohydrates, amino acids, organic acids, and polymers. The most often used carbon source for microbial processes is carbohydrates such as glucose, sucrose, and lactose since these compounds are easily metabolized by most industrially important filamentous fungi. The uptake of carbohydrates is normally restricted to monosaccharides, whereas disaccharides are degraded by hydrolytic enzymes excreted to the surrounding medium or present in the cellular membrane. Unlike the situation for bacteria, sugar transport into filamentous fungi does not seem to involve a phosphorylation step [21]. Free monosaccharides do not normally accumulate in the cell during uptake and most hexoses are converted to glucose-6-phosphate (G-6-P) or fructose-6-phosphate (F-6-P) before being metabolized. The intracellular isomerization of G-6-P to F-6-P is normally in equilibrium, and G-6-P can therefore be considered as a common end product in the sugar uptake reactions.

The catabolism of carbohydrates from G-6-P is traditionally divided into glycolysis and pyruvate metabolism. Glycolysis is defined as the sum of different pathways metabolizing G-6-P, and for filamentous fungi three pathways have been described [13]: 1) The Embden-Meyerhof-Parnas pathway (EMP), 2) the pentose phosphate pathway (PP), and 3) the Entner-Doudoroff pathway (ED). An overview of the three pathways is given by Berry [14]. The ED pathway has only been described in a few species of filamentous fungi (not including the industrially most important species) and it is therefore not discussed further here. The EMP and PP pathways serve two different purposes in the cellular metabolism. The major purpose of the EMP pathway is to generate energy and pyruvate, which may be further metabolized. The PP pathway is the major supply of NADPH, which is mainly used as reductive power in the biosynthetic reactions, and in addition the

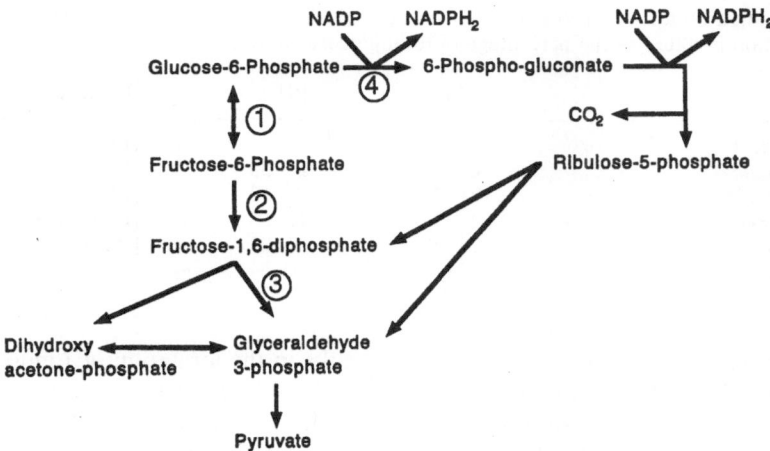

Fig. 2. A summary of the EMP and the PP pathways in filamentous fungi. The enzymes are *1*) Isomerase, *2*) phosphofructokinase, *3*) fructose-1,6-diphosphate aldolase, and *4*) glucose-6-phosphate dehydrogenase

pathway is important in providing building blocks in the form of ribose-5-phosphate, which is needed for the production of nucleotides. A summary of the EMP and the PP pathways is given in Fig. 2.

The distribution of glucose metabolism through the two pathways EMP and PP has been examined for several species, and the results for a few selected filamentous fungi are summarized in Table 1 (see Blumenthal [13] for a comprehensive list). Carter and Bull [16] have shown that the distribution of glucose metabolism through the two pathways in *Aspergillus nidulans* is determined by the operating conditions. In a glucose-limited chemostat at steady state the distribution of glucose through the PP pathway increased (not linearly) from 21% to 38% when the dilution rate was increased from 0.030 h^{-1} to 0.072 h^{-1}. The results are supported by a relatively higher increase in the activity of glucose-6-phosphate dehydrogenase (an enzyme in the PP pathway) than in the activity of fructose-1,6-diphosphate aldolase (an enzyme in the EMP pathway) with increasing dilution rate. The flux through the EMP pathway is controlled by a variety of intracellular components including ATP, which allosterically controls the enzyme phosphofructokinase (see Fig. 2). Thus at high glucose concentrations (high dilution rates in the chemostat) excessive flux through the EMP pathway is prevented at the level of conversion of F-6-P to fructose-1,6-diphosphate by the phosphofructokinase, and since the equilibrium constant of the isomerase is close to 1,F-6-P does not accumulate but it is converted to G-6-P which is channeled into the PP pathway [21].

Pyruvate formed in the glycolysis is a central compound in cellular metabolism. It may be oxidized completely to carbon dioxide and water in the tricarboxylic acid cycle (TCA cycle) through acetyl-CoA or it may be converted to acetate, ethanol, or lactate. The most energy efficient metabolism of pyruvate is the complete

Table 1. The distribution of glucose metabolism through the two pathways EMP and PP.
The distribution is stated as the percentage of total glucose consumed

Species	EMP	PP	Reference
Aspergillus niger	78		[15]
Aspergillus nidulans	62–79	21–38	[16]
Neurospora crassa	88–99	1–12	[13]
Penicillium chrysogenum	42–77	23–58	[17, 18, 19]
Penicillium digitatum	77–83	17–23	[19, 20]

oxidation to carbon dioxide and water, which results in the formation of 1 mole
ATP, 4 moles NADH, and 1 mole $FADH_2$ for each mole pyruvate oxidized. A
prerequisite for complete oxidation of pyruvate in the TCA cycle is that NAD^+
and FAD can be regenerated from NADH and $FADH_2$. This is done in the
oxidative phosphorylation which is an oxidative process involving free oxygen.
NADPH formed in the PP pathway may also be oxidized to $NADP^+$ in the oxidative phosphorylation, but this process is less energy efficient than the regeneration
of NAD^+. The reason is that NADPH formed from the PP pathway is present
in the cytoplasm and not in the mitochondria, and the compound therefore enters
the oxidative phosphorylation process at a different position than NADH [22]. It is
therefore energetically better suited to oxidize G-6-P completely to carbon dioxide
via the EMP pathway and the TCA cycle than via the PP pathway.

Most filamentous fungi are strict aerobes since they cannot generate sufficient
energy for growth by the conversion of pyruvate to acetate, ethanol or lactate
only. During submerged growth with low dissolved oxygen concentrations the
oxidative phosphorylation may become a bottleneck resulting in accumulation in
the medium of ethanol, lactate, acetate, or TCA cycle intermediates, e.g. citric
acid. Carter and Bull [16] also found that decreasing dissolved oxygen concentration resulted in an increased distribution of glucose into the PP pathway. For a
detailed description of the TCA cycle and the oxidative phosphorylation in fungi see
Niederpruem [23] and Lindenmayer [24].

As stated earlier a glucose limited chemostat is a good tool for obtaining
experimental data which may help to reveal how the complex carbon metabolism
discussed above varies with varying glucose concentration. It is, however, necessary
to measure all the major carbon containing compounds if solid conclusions are
to be made. To determine the distribution of glucose through the different pathways
it is necessary to apply more sophisticated techniques [13] (normally involving
carbon isotopes), but also here it is advantageous to use the steady state chemostat
as illustrated by Carter and Bull [16] (see discussion above). Unfortunately the
experimental program applied by most researchers only includes measurement of
the glucose and the biomass concentrations in the medium combined with
measurements of the oxygen uptake rate (OUR) and the carbon dioxide evolution
rate (CER). Through simple stoichiometric calculations [7] it is, however, often
possible to extract valuable information concerning the carbon metabolism from
these few measurements as illustrated in the following.

With growth on a simple medium where ammonia is the nitrogen source and glucose is the only carbon source we can summarize the overall growth process by the stoichiometric equation (5) (see Roels [7] for details in deriving the stoichiometric equation). The reaction is the sum of all intracellular reactions — including the energy forming reactions (e.g. the EMP and the PP pathways) and the reactions involved in formation of building blocks. The stoichiometry in Eq. (5) only holds if all carbon in the glucose is recovered either in the biomass or as carbon dioxide. With the formation of byproducts it is necessary to modify the stoichiometry, and with formation of only one byproduct it is given by Eq. (6), where γ_p is the degree of reduction of the byproduct [7]. The yield coefficients Y_{sx} and Y_{sp} in Eqs. (5) and (6) are given as C-moles per C-mole glucose, and they are not necessarily constants, i.e. they may vary with the operating conditions. For growth without product formation it is observed that the respiratory quotient (RQ) is always larger than 1. For growth with product formation RQ depends on the nature of the byproduct: $RQ > 1$ if $\gamma_p > 4$, but depending on the size of the yield coefficients RQ may attain values both larger and smaller than 1 if $\gamma_p < 4$. With ethanol as the byproduct ($\gamma_p = 6$) RQ is always larger than 1, which is a well-known observation for processes involving the yeast *Saccharomyces cerevisiae*. A comprehensive list of the degree of reduction for different metabolic products is given by Roels [7].

Growth without byproduct formation

$$CH_2O + \left(1 - \frac{4.2}{4} Y_{sx}\right) O_2 + 0.15 Y_{sx} NH_3 \longrightarrow Y_{sx} CH_{1.65}$$

$$\cdot O_{0.5} N_{0.15} + (1 - Y_{sx}) CO_2 + (1 - 0.6 Y_{sx}) H_2O \tag{5}$$

$$RQ = \frac{1 - Y_{sx}}{1 - 1.05 Y_{sx}}$$

Growth with byproduct formation

$$CH_2O + \left(-\frac{4.2}{4} Y_{sx} - \frac{\gamma_p}{4} Y_{sp}\right) O_2 + 0.15 Y_{sx} NH_3 \longrightarrow Y_{sx}$$

$$\cdot CH_{1.65} O_{0.5} N_{0.15} + Y_{sp}P + (1 - Y_{sx} - Y_{sp}) CO_2 + Y_{sw} H_2O \tag{6}$$

$$RQ = \frac{1 - Y_{sx} - Y_{sp}}{1 - 1.05 Y_{sx} - \gamma_p/4 \, Y_{sp}}$$

By combining measurements of the glucose and the biomass concentration in a steady state chemostat with measurements of the CER it is possible to set up a carbon mass balance, and this may indicate whether a byproduct is formed or not. If a byproduct is formed it is possible to calculate γ_p, provided measurements

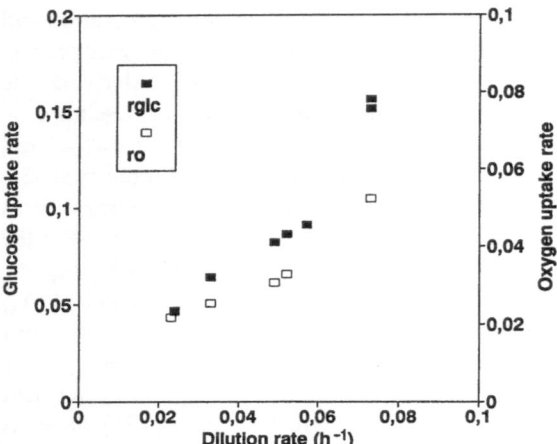

Fig. 3. Measurements of the glucose uptake rate and the OUR with varying dilution rate for *Aspergillus nidulans* in a steady state chemostat. The glucose uptake rate (r_{glc}, C-mole glucose per C-mole biomass per h) and the oxygen uptake rate (r_o, mole oxygen per C-mole biomass per h). The data are taken from Carter et al. [25] and they are converted to C-mole basis by assuming the biomass composition as stated in Eq. (5) and an ash content of 5%

of OUR are also made. In Fig. 3, experimental data are shown for a glucose-limited steady state chemostat with *Aspergillus nidulans* (the data are taken from Carter et al. [25]). It is observed that the glucose uptake rate and the OUR increase linearly with increasing dilution rates up to $0.055 \, h^{-1}$. This corresponds to constant values of Y_{sx} and Y_{ox} (respectively 0.74 and 2.95). It is observed that the glucose uptake rate and the OUR are larger than zero for $D = 0$ due to maintenance reactions. The RQ is approx. 1.1 for $D < 0.055 \, h^{-1}$, and from this value and the Y_{ox} the carbon recovery in the carbon dioxide is calculated to be 0.27. It may therefore be concluded that no byproduct is formed for $D < 0.055 \, h^{-1}$, and if we calculate RQ from Eq. (5) we find $RQ = 1.17$, which corresponds quite well to the measured value. At higher dilution rates there is a decrease in the yield coefficients, which indicates a shift in metabolism. This is supported by an observed increase in RQ to 1.35. For $D > 0.055 \, h^{-1}$ we find $Y_{sx} = 0.38$ and $Y_{ox} = 1.14$. For $RQ = 1.35$ it is calculated that the carbon recovery in the carbon dioxide is 0.45, and a byproduct must therefore be formed with $Y_{sp} = 0.17$. From Eq. (6) we calculate the degree of reduction of the 'missing' product to be 6.4, which indicates that P could be an alcohol, e.g. ethanol ($\gamma_{ethanol} = 6$). It is, however, not possible to make a firm conclusion on the basis of the few experimental data for $D > 0.055 \, h^{-1}$, but addition of ethanol measurements to the experimental program would have been a good idea. Furthermore more measurements should be made for $D > 0.055 \, h^{-1}$.

Formation of ethanol does not explain the decrease in Y_{ox} for high dilution rates (the observation is the complete opposite of that made during growth of *Saccharomyces cerevisiae* where Y_{ox} increases when ethanol production is initiated). It is, however, observed that the shift in metabolism is followed by an increase in the distribution of glucose through the PP pathway (see above and Carter and Bull [16]). Since the oxidation of glucose through the PP pathway gives less energy than oxidation of glucose through the EMP pathway and the TCA cycle, it is necessary with an increasing flux through the EMP pathway when the distribution of glucose to the PP pathway increases. This explains the increase in the specific glucose uptake rate at high D values, i.e. Y_{sx} decreases. The decrease in Y_{ox} indicates that the rate of the oxidative phosphorylation also increases, but if the flux through the TCA cycle increases less than the flux through the EMP pathway it is likely that some of the pyruvate is converted to a byproduct, e.g. ethanol.

A decreasing Y_{ox} with increasing dilution rate in a glucose limited chemostat has also been reported by Pirt and Callow [26] for *Penicillium chrysogenum* (two different strains examined). This could indicate that the hypothesis discussed above also holds for this species (which is in the same family as *Aspergillus nidulans*). The carbon recovery as biomass and carbon dioxide was approx. 100% except at one high dilution rate (D = 0.079 h^{-1}) where almost 20% of the glucose carbon is 'lost'. At this dilution rate $Y_{sx} = 0.497$, $Y_{sc} = 0.312$, and $Y_{ox} = 1.66$. With these values it can be calculated that $\gamma_p = 3.8$ for a missing product, which could indicate that gluconic acid is formed ($\gamma_{gluconate} = 3.7$). The carbon balance is, however, strongly dependent on the measurements of oxygen and carbon dioxide in the exhaust gas, and with the specified uncertainty in the RQ measurement (approx. 10%) it is not reasonable to make any firm conclusions from this experiment alone.

The two sets of experimental data discussed above indicate changes in the carbon metabolism for varying dilution rates in a glucose limited chemostat, and there is reasonable agreement between the results. Other experiments with filamentous fungi do, however, show diffeent trends. Pirt and Righelato [27] and Ryu and Hospodka [28] found no changes in Y_{sx} for varying dilution rates, whereas Mason and Righelato [29] found a downwards shift in Y_{sx} and an increase in Y_{ox} when the dilution rate was increased above 0.055 h^{-1} (both examining *Penicillium chrysogenum*). Mason and Righelato did, however, use a medium with sucrose as the carbon source, and it is likely that the carbon metabolism is different for growth on this sugar rather than on glucose.

For a complex nitrogen source (e.g. corn steep liquor) the stoichiometry is more complex than stated in Eqs. (5) and (6), but in principle it is possible to do the same type of analysis. It may, however, be difficult to specify a good overall composition of the nitrogen source. Despite the difficulties with quantification of a complex nitrogen source one should always check the mass balances carefully, and if there is some 'missing' carbon one may apply stoichiometric calculations to guide the search for the 'missing' product.

With the complex carbon metabolism of filamentous fungi it is obviously not possible to apply a simple unstructured growth model, e.g. a Monod model, where all rates are taken to be linearly dependent of the specific growth rate. In the past there have been no attempts to set up good mechanistic models for the catabolism of filamentous fungi, but it is likely that much of the modelling effort made in the area of yeast growth [5, 30] may be used with success.

3.2 Biosynthesis

Filamentous fungi have a large biosynthetic machinery as illustrated by their ability to utilize ammonia as the sole nitrogen source. All major building blocks for macromolecular synthesis can be synthesized from ammonia and glucose (or another sugar), and filamentous fungi may therefore grow on a simple synthetic medium. However, in industrial processes a combination of ammonia and a complex nitrogen source (e.g. corn steep liquor or yeast extract) is normally used, since the availability of proteins and amino acids in the medium gives enhanced growth and product formation of many species.

Experimental results indicate that ammonia is taken up in the form of NH_4^+ by a specific transport system [31, 32], and the process must be coupled with an excretion of cations or a parallel uptake of anions to maintain electroneutrality. Roos and Luckner [33] suggest that electroneutrality is maintained by excretion of an undissociated carboxylic acid formed in the catabolism, e.g. citric acid. Since a net formation of carboxylic acids has not been reported in glucose-limited chemostat cultures at low dilution rates — even when ammonia is the only nitrogen source — the ammonia uptake must be able to couple with the transmembrane transport of other compounds. The ammonia uptake system may transport ammonia against a concentration gradient resulting in significantly higher intra-cellular than extracellular concentrations, and the system is feed-back controlled by L-glutamine, which plays a central role in the nitrogen metabolism [31].

The uptake of amino acids in filamentous fungi has been examined by Hunter and Segel [34, 35]. A combination of highly specific and less specific transport systems ensure that the right amounts of different amino acids are taken up by the cell. Some of the transport systems are constitutive whereas other are regulated by various factors, e.g. the transport of L-methionine and L-cysteine is regulated by the availability of sulfur [32].

The amino acid biosynthesis in filamentous fungi resembles that found in yeast and bacteria [32], and precursors for the biosynthesis are intermediates in the EMP pathway and in the TCA cycle. The anaplerotic pathways compensate for the removal in the TCA cycle of the intermediates which are used for biosynthesis of amino acids. These pathways thereby ensure preservation of the energy generating function of the TCA cycle [36]. In filamentous fungi, anaplerotic metabolism is provided by: 1) The glyoxylate cycle, 2) acetyl-CoA synthetase activity, and 3) direct fixation of carbon dioxide [21]. The glyoxylate cycle enables the decarboxylating reactions of the TCA cycle to be bypassed resulting in a net synthesis of succinate. The pathway is active when growth is sustained by acetate (and perhaps other carboxylic acids) and is usually inactive with growth on carbohydrates. Carbon dioxide fixation is mainly carried out by pyruvate carboxylase which catalyzes the conversion of pyruvate to oxaloacetate. Another important enzyme is phosphoenolpyruvate carboxykinase, which ca-talyzes the decarboxylation of oxaloacetate with formation of phosphoenol-pyruvate, which may ensure formation of pyruvate with growth on acetate and other carboxylic acids. For a review of anaplerotic pathways one is referred to [36].

Like many other microorganisms filamentous fungi have an intracellular pool of free amino acids, and for several species it is found that the level of L-glutamate is higher than that of any other amino acid-something which may be explained by the central role of this amino acid in the biosynthesis reactions, i.e. it is a precursor for the formation of several other amino acids. Besides being used for synthesis of proteins the amino acids are precursors for purines and pyrimidines, which are building blocks for RNA and DNA. Furthermore, amino acids are precursors for a wide range of secondary metabolites, e.g. penicillin, and it is therefore obvious that the size of the intracellular amino acid pool is an important factor for growth and product formation. In the past there have, however, not been any attempts to set up structured models for filamentous fungi which include the intracellular amino acid pool as a variable, but it is likely that such models will appear in the future. Here one could probably benefit from the existing detailed models for *Saccharomyces cerevisiae* [37].

3.3 Macromolecular Composition

Few experimental data are available for the macromolecular composition in filamentous fungi grown at vaious environmental conditions, and the mechanisms behind cell growth are not as well known as for prokaryotes [12]. This is a major obstacle for setting up good structured models, since vertification by comparison with measurements of key intracellular variables, e.g. the ribosome level or the ribosomal RNA content [38], is difficult.

Studies of *Neurospora crassa* revealed a linear relation between the specific growth rate and the level of ribosome activity [39, 40] − an observation also made for bacteria [11]. The ribosomes are responsible for the protein synthesis which are the most energy demanding processes in the cell, and it is therefore reasonable that the cells adjust the level of ribosomes to the environmental conditions, i.e. the specific growth rate. Measurements in *N. crassa* at high specific growth rates show that the ribosome level (number of ribosomes relative to the amount of amino acids bound in protein) increases with the specific growth rate whereas the average ribosome efficiency (amino acid polymerized per ribosome per minute) is almost constant [39]. At low specific growth rates the ribosome efficiency is found to decrease whereas the ribosome level is constant, an observation also made with *Aspergillus nidulans* [41] and bacteria [42]. This indicates that the cells have some unused capacity at low specific growth rates − a capacity which may be rapidly activated when changes in the environmental conditions occur. In a simple structured model one should therefore include both the ribosome efficiency and the ribosome level, if the ribosomes are used to represent the cellular activity.

The ribosomes consist of 60% RNA − normally referred to as ribosomal RNA (or rRNA) − and the rest being mainly protein. Most of the stable RNA in the cell is rRNA and one may therefore easily obtain a good estimate of the ribosome level from measurements of the intracellular RNA content. For bacteria a large number of data for the RNA content are available, but unfortunately only few measurements of the RNA content in filamentous fungi have been reported. Table 2 summarizes some of the data for *Penicillium chrysogenum*. It is observed that the RNA content increases with the specific growth rate and for a very low specific growth rate $(0.003 \ h^{-1})$ the RNA content is larger than zero. This corresponds with the results discussed above.

The central role of the ribosomes in the cellular growth is supported by experiments with transient operating conditions where a rapid adjustment of the rRNA to the new environmental conditions is observed [39]. During a shift down in the specific growth rate it is observed that the net synthesis of rRNA stops for a period, and when the level corresponding to the new specific growth rate is obtained the rRNA synthesis is resumed [40]. The exact control mechanisms for synthesis of rRNA are not known but it is likely that there are similarities to the mechanisms found in bacteria [11, 42].

Measurement of other intracellular macromolecules such as DNA, protein, and carbohydrates is also reported in the literature [39, 40, 44]. Measurement of rRNA is, however, the most interesting from a modelling point of view due to the key role of the ribosomes.

Table 2. Measurements of the RNA content in *Penicillium chrysogenum* at different specific growth rates. The last two data points are results from our own laboratory

Conditions	Specific growth rate (h^{-1})	RNA content per biomass $(kg\ kg^{-1})$
Sucrose limited chemostat	0.003	0.020
[43]	0.043	0.046
	0.110	0.072
Glucose limited chemostat	0.030	0.055
	0.050	0.053

4 Growth Mechanisms of Filamentous Fungi

Filamentous fungi are multicellular microorganisms and their growth is significantly different from that of unicellular microorganisms. The individual cells are connected in so-called hyphal elements or mycelia, which arise from the outgrowth of a single cell or spore. Upon spore germination the growth of the spore rapidly becomes polarized resulting in formation of a thread, or hypha, which grows only at the tip. During growth of the hypha new tips are formed along the thread by a mechanism called branching. Along the hypha there may be functional differentiation, e.g. different processes run in "tip cells" and in "older cells" behind the tips. The differentiation processes have a significant effect on the growth potential of the fungi, and they must be considered in any good growth model. Furthermore, production of secondary metabolites often have a relationship to the differentiation processes, and a good growth model may therefore easily be extended to describe product formation.

4.1 Spore Germination and Sporulation

The fungal spore can be considered as the beginning and the end of the development cycle of a fungi. The spore is the dormant state of the organism, since there is no synthesis of cellular material and metabolic activity is much reduced. Spores are characterized by a minimal metabolic turnover, low water content, and lack of cytoplasmic movement. Normally one distinguishes between two types of dormancy: constitutional and exogenous [45]. Germination of spores in constitutional dormancy requires an activation process to overcome a specific property of the dormant state, i.e. the spores do not germinate simply by placing them under suitable environmental conditions. Spores in exogenous dormancy are inactive simply because of unfavourable chemical and/or physical conditions and they quickly resume growth in a suitable environment.

Spore germination normally requires the presence of several substrate components (carbon source, nitrogen source, oxygen etc.) together with the correct physical conditions (moisture level, temperature etc.). Often, the spore germination requires a more complex medium than necessary for vegetative growth, i.e. species that grow on a synthetic medium may not necessarily germinate on the same medium. Prior to germination the spore increases in size, and there is a rapid increase in the number of mitochondria and in the content of the active cell material, e.g. the size of the endoplasmatic reticulum (location of the ribosomes). Initially during germination, the spore grows as a sphere and new cell material is laid down uniformly over the entire inner surface of the spore. At a given time, growth polarity is established and wall formation ceases at all but one two areas which grow out resulting in formation of a hyphal element. The outgrowth of a hypha has been proposed to occur by the same mechanisms as those involved in hyphal tip extension [45, 46], see Sect. 4.2.

For industrially important species of *Aspergillus* and *Penicillium*, sporulation is rarely observed in submerged cultures, probably due to many factors, e.g. the physical nature of the hyphal wall during submerged growth makes sporulation difficult and the homogeneous nature of the medium ensures a direct contact of the spore forming cells with nutrients [47]. Chemostat experiments indicate that the C/N ratio is a major determining factor for sporulation, and that sporulation mainly occurs during nitrogen limitation. Furthermore, sporulation is primarily observed at low specific growth rates [47]. With *Penicillium chrysogenum*, occurrence of spores was observed at very low dilution rates even in a glucose limited chemostat [44]. When the glucose feed rate was lowered to the maintenance demand or a little above this level an increase in the number of spores was observed. At the same time the RQ decreased and acid began to accumulate in the medium. This indicates that glucose was not the limiting substrate component, i.e. the culture could have become nitrogen limited during the experiment, and changes in the C/N ratio could explain the increased sporulation.

4.2 Hyphal Tip Extension

The growth of hyphal elements takes place only at the tips, which extend linearly at a rate depending on the hyphal composition and the environmental conditions. Several cells behind the tip are involved in the tip extension process since they supply the necessary cellular material for tip extension, e.g. cytoplasmic material and building blocks for wall synthesis. Just behind the tip there are several cells which are not separated by cell walls (so called septa), and these cells therefore have a common cytoplasm in which the nucleus of each cell is positioned. The part of the hyphal element between the apex and the first septum is normally referred to as the apical compartment. Cells positioned right behind the apical compartment have an intracellular composition very similar to that of cells in the apical compartment, and this part of the hypha is referred to as the subapical compartment (one may also find the term intercalary compartment used in the literature). Despite the presence of septa there may be an exchange of protoplasm between the subapical and the apical compartment since the septa are normally perforated. With growth on solid media the cellular structuring begins to change and large vacuoles appear in the cells when one moves further away from the tip. The vacuoles in cells close to the subapical compartment are small, but they become larger when the distance to the tip increases. One of the functions of the vacuoles may be to create a sufficient intracellular pressure which ensures transport of protoplasm towards the tip section.

Many of the observations made with growth mechanisms of filamentous fungi are made with surface cultures, where the differentiation processes are more distinct than in submerged cultures [48]. When fungal spores germinate on a solid medium narrow hyphae with short apical compartments are formed. When the culture grows it gradually develops into a so-called "mature mycelium", where the hyphal element is differentiated into a leading hypha and narrower branch hyphae [48]. The leading hypha differs in diameter and apical compartment length from the primary and secondary branches on the hypha. The leading hypha has a very high tip extension rate compared with that found in an undifferentiated mycelium. The exact mechanisms behind the differentiation into a "mature mycelium" are not known, but there is most likely an influence of the substrate which may be depleted in certain zones on the solid medium where the culture is dense. An undifferentiated culture is produced under the constant or near constant conditions which prevail during exponential growth, and it has been shown that the morphology of undifferentiated hyphal elements grown on solid media is very similar to the morphology of hyphal elements in a submerged culture [48]. To model the growth of submerged cultures one may therefore use experimental data from undifferentiated cultures on solid media. Many of the mechanisms described in the following have been discovered through studies of cultures grown on solid media.

Since cell division does not occur in filamentous fungi it is difficult to define one cell cycle. Fiddy and Trinci [49] introduced the term duplication cycle to describe events during fungal growth which are analogous to those observed during the cell cycle of unicellular microorganisms. The main events in the duplication cycle are [50]: 1) A newly formed apical compartment increases in length at a liner rate, 2) the volume of cytoplasm per nucleus increases and when it reaches a critical ratio synthesis of nuclei (or chromosomes) is initiated, 3) the synthesis of nuclei is exponential until the number of nuclei has doubled, 4) when the apical compartment has reached a size corresponding to twice the initial size a septum is formed in the middle of the compartment, and a "new" apical compartment has been formed. The duration of the duplication cycle has been found to be identical with the doubling time of the biomass for *Aspergillus nidulans* [49]. Despite the similarities with the cell cycle of unicellular microorganisms, the duplication cycle does not represent the formation of one new cell but a whole new apical compartment which may comprise several cells.

Electron microscope studies of filamentous fungi revealed that there is a large accumulation of small vesicles at the apex [51], and it is believed that these vesicles play a central role in the tip extension. Bartnicki-Garcia [52] proposed that wall subunits, lytic enzymes, and synthetic enzymes are incorporated into vesicles at specialized regions of the endomembrane system in the apical and subapical compartments. The vesicles, each carrying its load of enzymes and/or wall precursors, are then transported through the cytoplasm to the tip section of the apicall cell-normally referred to as the extension zone [51]. When a vesicle comes in contact with the cell membrane at the apex it fuses with the membrane and the vesicle content is excreted into the wall region at the apex. The excreted lytic enzymes attack the microfibrillar skeleton in the cell wall resulting in a weakened complex which is not able to withstand the inner pressure from the cytoplasm. The microfibrils therefore become stretched and less integrated, and the surface area of the wall increases. The dissociated microfibrils are then rebuilt from the precursors by the synthetic enzymes, and the cell wall has expanded one unit area without losing its overall properties [45, 52].

The early electron microscope studies on hyphal tips disclosed the existence of two vesicle populations of different sizes, i.e. macro- and microvesicles, but it was only recently verified that the two types of vesicles serve entirely different functions during tip extension [46]. The microvesicles have been isolated and they are found to contain mainly chitin synthetase − the enzyme necessary for synthesis of the characteristic cell wall in fungi. These vesicles are therefore referred to as chitosomes. Despite their larger size the macrovesicles have not been isolated in a reasonably pure form, and their exact biochemical composition and function is therefore not known. It is, however, believed that these vesicles provide the necessary building blocks for tip extension and the lytic enzymes necessary for plasticizing the wall [46]. The macrovesicles are also believed to be the vehicle for secretion of extracellular enzymes. With two vesicles involved in tip extension the original model of Bartnicki-Garcia [52] described above is slightly modified [46].

It has been estimated that about 38,000 vesicles fuse with a tip of a *Neurospora crassa* hypha per min during tip extension [51], and a constant supply of macrovesicles to the extension zone is therefore strictly necessary to maintain growth. Macrovesicles are synthesized in both the apical and in the subapical compartments and the synthesis probably takes place while they are transported towards the apex. The microvesicles are synthesized at or near the plasma membrane-cell wall interface [46], i.e. the position where they are to be used. The mechanisms that govern transport of macrovesicles to the apex are not known, but the continuous production of protoplasm throughout the hyphal element results in formation of an intracellular pressure gradient, which may partly explain the transport of material. This transport by protoplasmic streaming does, however, not explain the observed accumulation of vesicles at the tips.

The mechanisms of vesicle supply to the apex has been examined by Bartnicki-Garcia using a computer model [46]. The model is based on an assumption of continuous delivery of vesicles at the same rate in all directions from a *vesicle supply center* (VSC). The VSC is not the place of synthesis of vesicles but a position in the hypha where the vesicles are collected before they are distributed to the apex. If the VSC is stationary in the center of a sphere cell growth is obviously spherical, but if the VSC is displaced while continuing the release of vesicles, different forms will be generated and the final shape depends on the direction and the velocity of displacement of the VSC. By displacing the VSC linearly from the center of the initially spherical cell, the computer model predicts the formation of a hypha.

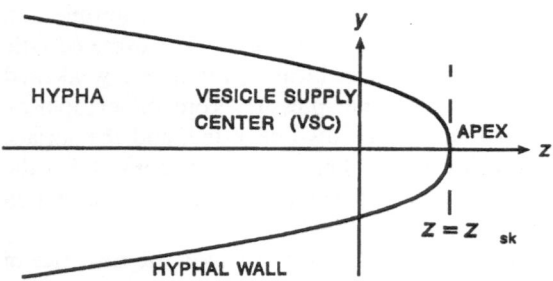

Fig. 4. Shape of ideal hypha calculated from (7) with $z_{sk} = 0.80\ \mu m$, corresponding to a maximum hyphal diameter of $D = 5\ \mu m$

Examination of hyphal profiles for a large number of different species of filamentous fungi (obtained from electron microscopy) revealed that in two dimensions, the shape of the hyphal tube can be described by Eq. (7), where y and z are the radial and axial directions of the hypha, and the apex is positioned at $(z_{sk}, 0)$ [46], see Fig. 4.

$$z = y \cot\left(\frac{y}{z_{sk}}\right) \tag{7}$$

Equation (7) describes the profile of a hypha in its entirely and it therefore differs significantly from other mathematical expressions found in the literature which are mainly approximations of the apical dome [53]. It is obvious that Eq. (7) only describes a mathematical shape which to a certain approximation resembles that of a real hypha [46]. The model parameter z_{sk} determines the shape of the hypha, and it represents a coupling to the computer model described above. If z_{sk} is given by Eq. (8) — where r_v is the number of vesicles released from the VSC per unit time (equal to the rate of increase in the wall area, $m^2 h^{-1}$) and r_{VSC} is the rate of linear displacement of the VSC — it represents the distance from the VSC to the apex, i.e. the VSC is positioned at (0,0), see Fig. 4. One could expect that r_{VSC} is determined by the tip extension rate ($m h^{-1}$) whereas r_v depends on the level of vesicles at the VSC (which again depends on the rate of supply of vesicles to the VSC).

$$z_{sk} = \frac{r_V}{r_{VSC}} \qquad (8)$$

From Eq. (7) it is easily shown that the maximum diameter (d_{max}) of an ideal hypha is given by $2\pi z_{sk}$.

The calculated position of the VSC in the hypha may be physiological significant, since it was found that it coincides with the position where a large number of vesicles seem to aggregate — the so called Spitzenkörper [54]. The exact function of the Spitzenkörper is not known, but it is believed to be a center for the final distribution of vesicles. Extension of the model to include branching will give it an even wider application for examination of various hypotheses concerning the growth mechanisms in filamentous fungi. The model cannot be used to predict growth of whole cultures since the influence of the environment is not included and the mechanisms for the VSC displacement are not described explicitly.

4.3 Branching

Since a hyphal element grows only at the tips the rate of biomass formation (kg per m^3 per h) is given by Eq. (9), where ε is the "concentration" of hyphal elements in the culture (m^{-3}), n is the average number of actively growing tips in the hyphal elements, and r_{tip} is the average tip extension rate (kg per tip per h).

$$r_x = \varepsilon n r_{tip} = \mu x \qquad (9)$$

r_{tip} has a certain maximum value [55], and if ε is fairly constant (see discussion in Sect. 6.1) it is obvious that exponential growth, i.e. μ is constant, can only be sustained for many generations if n increases, i.e. new tips are formed along the hypha. This is ensured by a mechanism called branching.

There seems to be certain "preferred" branching points on a hypha. Trinci [50] suggests that branches are initiated at locations where for one reason or another there is an accumulation of vesicles, i.e. at the "preferred" branching points. Since vesicles are produced throughout the hypha at a constant rate it seems reasonable that at positions where the protoplasmic flow is reduced, e.g. at the position of a septum, there will be an accumulation of vesicles. This indicates that branching is associated with the formation of septa [56], and for *Geotrichum candidum* more than 70% of the observed branch points in a subapical compartment are positioned close to the septum separating this compartment from the apical compartment [57]. For *Aspergillus nidulans* the distribution of branch points is much more equal throughout the subapical compartment. This difference may be explained by the very small pores in septa in *Geotrichum candidum* whereas *Aspergillus nidulans* has large central septal pores through which vesicles may easily pass [57]. In most cases branching occurs in the subapical compartment, but in some cases filamentous fungi may branch apically. This event occurs when the rate of supply of vesicles to the apex exceeds the rate at which they can be incorporated into the wall of the extension zone [56]. Little is known of the mechanisms behind branching but new experiments with *Fusarium graminearum* indicates an effect of cAMP and cGMP [58], which are central components in the regulation of many different intracellular events. With a high cAMP concentration the hyphal elements become more densely branched whereas the effect of cGMP is the opposite [58]. The concentration of cAMP is normally low in conditions with excess substrate (e.g. high glucose concentrations), and the results therefore indicate that the branching frequency increases with low substrate concentrations. This contradicts with other investigations using *Geotrichum candidum* [59] where the hyphal elements are more densely branched at high glucose concentrations. From an evolutionary point of view one would expect that the branching frequency decreases with low substrate concentrations, since the fungi will try to direct growth away from environments where the substrate is limiting, and this is best done by forming a long hypha without branches. The applied strain of *Fusarium graminearum* is a highly branched mutant of a wild type strain, and this may explain the reported results [58].

For characterization of the morphology of hyphal elements Caldwell and Trinci [60] introduced the hyphal growth unit length (l_{hgu}) defined by the total length of the hyphal element divided by the number of tips. The hyphal growth unit is a valuable quantity for interpretation of experimental data, and it was originally considered as an analogue to the single cell in uncellular cultures. The hyphal growth unit length specifies the average length of a hyphal element involved in the growth of each tip, i.e. the length supporting vesicles necessary for tip extension. In order to account for varying hyphal diameters it is better to use the hyphal growth unit volume (V_{hgu}), which is the total volume of the hyphal element divided by the number of tips. It describes the average volume of cytoplasm involved in the growth of each tip, and it is therefore a better analogue to the single cell in unicellular cultures than l_{hgu}. In models for microbial systems one normally uses dry mass as the characteristic variable for the biomass, and for modelling the morphology of filamentous fungi it is therefore an advantage to introduce the hyphal growth unit mass (x_{hgu}), which is the average dry mass of a hyphal element divided by the number of tips. If one knows the average hyphal diameter (d), the density of the hypha (ϱ), and its water content (w) one may easily convert the different hyphal growth unit definitions to each other as illustrated in Eq. (10).

$$x_{hgu} = \varrho(1 - w) V_{hgu} = \varrho(1 - w) \frac{\pi}{4} d^2 l_{hgu} \qquad (10)$$

The density and the water content are normally constant whereas the hyphal diameter may vary with the environmental conditions.

From Eq. (9) one finds:

$$r_{tip} = \mu \, \frac{x}{\varepsilon n} = \mu x_{hgu} \tag{11}$$

which shows that if the hyphal growth unit mass is constant there is direct proportionality between the average tip extension rate and the specific growth rate. A similar correlation is reported in the literature for hyphal growth unit length [48] where r_{tip} has the unit (m hypha per tip per h).

If new tips are formed only by branching, i.e. ε is constant, the dynamic balance for the number of tips is given by Eq. (12) where φ is the branching frequency (tips per kg DW per h).

$$\varepsilon \, \frac{dn}{dt} = \varphi x \tag{12}$$

by using $x_{hgu} = x/(n\varepsilon)$ and Eq. (12) a dynamic mass balance for the hyphal growth unit mass may be derived:

$$\frac{dx_{hgu}}{dt} = x_{hgu}(\mu - \varphi x_{hgu}) \tag{13}$$

With a constant hyphal growth unit mass the specific growth rate equals the branching frequency times the hyphal growth unit mass. Equations (10), (11), and (13) are valuable when experimental data for the hyphal growth unit under varying environmental conditions are to be interpreted, especially if information concerning the tip extension rate and the branching frequency is to be extracted.

Examination of the morphology of filamentous fungi under varying environmental conditions shows that the hyphal growth unit length decreases and the hypha diameter increases with the specific growth rate [59, 61, 62]. A similar observation is made with different species of *Streptomyces* [63, 64]. Figure 5a shows data from growth of *Geotrichum candidum* in a glucose limited chemostat [59], and it is observed that the hyphal growth unit volume is approximately constant ($V_{hgu} \approx 1510 \, \mu m^3$) for different specific growth rates (equal to the dilution rate D in the chemostat). The tip extension rate and the branching frequency can therefore be calculated from Eqs. (11) and (13) and the results are shown in Fig. 5b. A constant V_{hgu} indicates that the cytoplasmic volume necessary for supporting the growth of one tip is independent of the specific growth rate, and the specific vesicle production must therefore increase with D. The reason may be that the fractional volume of the apical and the subapical compartment increases with D, i.e. a larger part of the hyphal element is involved in the growth process at high specific growth rates and only a small part of the hyphal element is "inactive" biomass (vacuolated cells). The total volume of the apical and subapical compart-

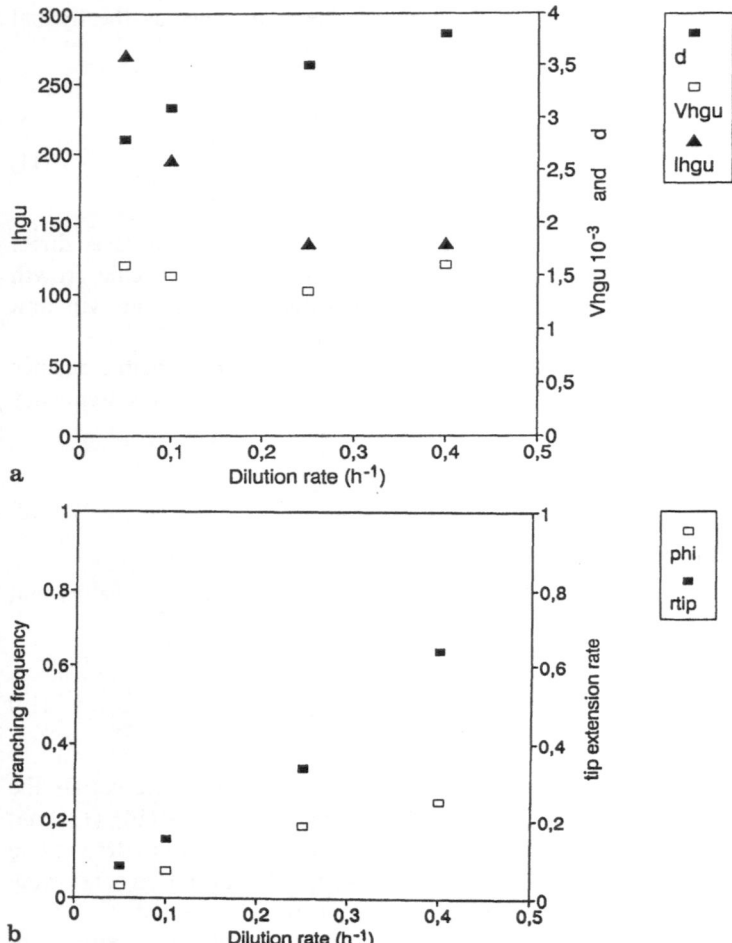

Fig. 5. Morphological variables at varying dilution rates in a glucose limited chemostat. The experimental data are taken from Robinson and Smith [59]. **a)** Hyphal growth unit length (μm), hyphal growth unit volume (μm³), and hyphal diameter (μm) as functions of the dilution rate. The hyphal diameter was found to be approximately the same for apical and subapical cells. **b)** Tip extension rate (μm³ per tip per h) and branching frequency (tip per μm³ per h) as function of the dilution rate

ments were observed to increase with D, but the total volume of the hyphal element is not reported [59]. The presence of many vacuolated cells in the hyphal elements at low dilution rates may be explained both by the low glucose concentration and the high residence time for the hyphal elements in the chemostat. It is impossible to discriminate between these two effects from the reported experiments. The increasing hyphal diameter with D may be caused by an amplified rate of vesicle production which results in an enlarged number of vesicles released from the VSC, i.e. r_v in Eq. (8) increases. Furthermore, the high vesicle production may result in an increased level of vesicles throughout the hypha and thus may give an increased branching frequency.

With V_{hgu} being approximately constant for varying environmental conditions, e.g. different glucose concentrations in the chemostat, one could expect that it is a characteristic parameter for a filamentous microorganism. In Table 3, values of V_{hgu} determined for different filamentous microorganisms are collected. A general observation for all the species is that V_{hgu} is fairly constant with varying environmental conditions. Furthermore, for species in the same family (*Aspergillus* and *Penicillium*) V_{hgu} is of the same size-especially when one considers the difficulties in determining both the hyphal growth unit length and the hyphal diameter precisely. For *Fusarium graminearum*, V_{hgu} increases rapidly with the specific growth rate [66] (both the hyphal growth unit length and the hyphal diameter increase), an observation which deviates from the general trend. At present, no reasonable explanation can be given for this, other than this fungi is known to exhibit a completely different branching pattern than the other species.

Data reported by Metz et al. [65] for *Penicillium chrysogenum* show that the hyphal growth unit length and hyphal diameter are approximately constant with varying dilution rates in a glucose limited chemostat, i.e. $l_{hgu} \approx 75\,\mu m$ and $d \approx 4.0\,\mu m$. This deviates from the general observation that the hyphal diameter increases with the specific growth rate. The experiments described by Metz et al. were carried out in a continuous bioreactor with a relatively high agitation rate

Table 3. Measurements of hyphal growth unit volume for different species of filamentous microorganisms

Species	V_{hgu} (μm^3)	Conditions	Ref.
Penicillium chrysogenum *T14*	1425	Shake flask, simple medium. $\mu = 0.14\,h^{-1}$	[62]
	1195	Glucose limited chemostat, $D = 0.12\,h^{-1}$	[62]
Penicillium chrysogenum *Q176*	1245	Shake flask, $\mu = 0.13\,h^{-1}$	[62]
Penicillium chrysogenum	~950	Glucose limited chemostat, varying D	[65]
Geotrichum candidum	~1510	Glucose limited chemostat, varying D	[59]
Aspergillus nidulans	1638	Shake flask, $\mu = 0.42\,h^{-1}$	[57]
Fusarium graminearum	230–1250	Glucose limited chemostat, varying D	[66]
Neurospora crassa spco 12	629	Shake flask, $\mu = 0.20\,h^{-1}$	[57]
Candidum albicans	217	Shake flask, $\mu = 0.39\,h^{-1}$	[57]
Streptomyces coelicolor	26	Shake flask, $\mu = 0.26\,h^{-1}$	[57]
Streptomyces hygroscopicus	1.3	Glucose limited chemostat, varying D	[64]

(1000 rpm), and the deviation from the general trend may be explained by a high degree of fragmentation, i.e. hyphal break up. The effect of fragmentation on hyphal morphology is discussed in Sect. 6.1.

5 Modelling the Growth Kinetics of Filamentous Fungi

Considering the industrial importance of filamentous fungi there has been remarkable few attempts to set up good growth models for these microorganisms. Very often, an unstructured model is applied to describe the growth process and product formation is included in an empirical fashion. If the applied model fails to describe all the experimental data it is often patched up with correction factors, e.g. by including an influence of a metabolic product, and the resulting model normally has very little predictive strength. A frequently applied model for penicillin production is that of Bajpai and Reuss [67], which describes the specific growth rate of the biomass by Contois kinetics (see Eq. (14)). When the model was used to simulate fed-batch operation it was necessary to adjust a key parameter in the model after a certain period of time [67]. In a preliminary study of a process, for which the experimental data are scarce, an empirical model may be used to examine different modes of operations, e.g. different feeding strategies during fed-batch operation [68]. When more knowledge is obtained one should, however, discard the empirical models and set up mechanistic models which may be used to verify hypotheses of the growth mechanisms or indicate how new experiments should be made to reveal new aspects of the rich life processes of filamentous fungi.

$$\mu = k \frac{s}{s + Kx} \qquad (14)$$

All the structured models reported in the literature are concerned with the morphological differentiation of the individual cells in the hyphal elements, and there have been no attempts to set up mechanistic models for the physiology of filamentous fungi, e.g. describe the catabolism discussed in Sect. 3.1. The models to be discussed here will therefore emphasize on the mechanisms behind morphological differentiation, and only structured models are considered. Most of the models include product formation, but since the growth process is the main topic of the present review, details of the product formation kinetics are not given. The models are presented within the general framework described in Sect. 2 (see also [5]).

5.1 Morphologically Structured Models

The model of Megee et al. [69] is the first model where the concept of morphologically structuring is used. Five different morphological forms are considered: *A — apical compartment in actively growing hyphae, H — subapical compartment in actively growing hyphae, C — conidiophore developing hyphae, B — black spores, and M — matured spores.* With these five morphological forms

Table 4. The model of Megee et al. [69]

Growth reactions

$$\alpha_1 S + Z_A \rightarrow Z_A + Z_H \qquad r_1 = k_1 \frac{S}{S + K_1} Z_A \qquad \text{tip extension} \qquad (15)$$

$$\alpha_2 S \rightarrow Z_H \qquad r_2 = k_2 \frac{S}{S + K_2} Z_H \qquad \text{assimilation} \qquad (16)$$

Metamorphosis reactions

$$Z_H \rightarrow Z_A \qquad u_1 = k_3 \frac{S}{S + K_3} Z_H \qquad \text{branching} \qquad (17)$$

$$Z_A \rightarrow Z_H \qquad u_2 = k_4 \frac{1}{S + K_4} Z_A \qquad \text{differentiation} \qquad (18)$$

---

it is possible to describe a complete life cycle of imperfect fungi (fungi with no sexual reproduction): growth of the hyphal elements, development of conidiophores, spore formation, and maturation of spores. For simulation of submerged growth one may neglect spore formation (see Sect. 4.1), and only consider actively growing hyphae, i.e. the morphological forms A and H. Hereby the original model is simplified substantially. The simplified version of the model is summarized in Table 4. Formation of cytoplasm from intracellular substrate S in the apical and subapical compartments is described by the reactions (15) and (16) respectively. Cytoplasm formed in Z_A is used for tip extension. If the size of the apical compartment is constant there is no net formation of Z_A upon tip extension, and (15) therefore describes a net formation of Z_H. Branching is described by the metamorphosis reaction (16) where Z_H is converted to Z_A, and differentiation is described by the metamorphosis reaction (17). Branching is positively influenced by the limiting substrate whereas differentiation is inhibited by high concentrations of the limiting substrate.

The specific growth rate of the whole biomass is given by Eq. (19) and since r_1 and r_2 specifies the rate of formation of similar material in the two morphological forms A and H one may expect that $K_1 \approx K_2$ and $k_1 \approx k_2$. Since $Z_A + Z_H = 1$ the expression for the specific growth rate simplies to Monod kinetics in the intracellular substrate concentration S (or as shown in [5] in the extracellular substrate concentration).

$$\mu = r_1 + r_2 = k_1 \frac{S}{S + K_1} Z_A + k_2 \frac{S}{S + K_2} Z_H \approx k_1 \frac{S}{S + K_1} \qquad (19)$$

The mass balance for Z_A is calculated from (4):

$$\frac{dZ_A}{dt} = u_1 - u_2 + (\mu_A - \mu) Z_A$$

$$= k_3 \frac{S}{S + K_3} Z_H - k_4 \frac{1}{S + K_4} Z_A - \mu Z_A \qquad (20)$$

where μ_A is the specific growth rate of the apical cells, which is zero since there is no net formation of Z_A in Eqs. (15) and (16). At the steady state one obtains:

$$Z_A = k_3 \frac{SZ_H}{S + K_3} \left(k_4 \frac{1}{S + K_4} + k_1 \frac{S}{S + K_1} \right)^{-1} \tag{21}$$

by tuning the model parameters it is possible to predict an increasing level of Z_A for increasing substrate concentrations. According to the discussion in Sect. 4.3 this is biologically reasonable. With ξ_A being the size of the apical compartment (kg (compartment)$^{-1}$) one may calculate the hyphal growth unit mass from:

$$x_{hgu} = \frac{\xi_A}{Z_A} \tag{22}$$

If ξ_A is constant the model predicts a decreasing x_{hgu} with increasing S in contradiction to the experimental data discussed in Sect. 4.3. Before the model can be used to simulate the morphology of hyphal elements it is necessary to describe ξ_A explicitly, e.g. as an empirical function of S.

In the original model of Megee et al. [69] formation of both primary and secondary metabolites are described. The primary metabolites are formed as byproducts of the growth process, i.e. reaction (16), whereas the secondary metabolites are formed as byproducts of the differentiation processes, i.e. reaction (18). The model describes several general observations made during growth of *Aspergillus awamori*, but a direct comparison with experimental data has not been made. A disadvantage of the original model is the large number of parameters, but the revised version of Table 4 has no more parameters than can be estimated from the experimental data discussed in Sect. 4.3. The revised version of the model − perhaps with product formation included − represents a simple structured model that may have some predictive strength.

Kristiansen and Sinclair [70] describe a morphologically structured model to be used for the simulation of citric acid production by *Aspergillus niger*. Three different morphological forms are considered: 1) Basic cells produced during nitrogen-unlimited growth, 2) storage cells formed from the basic cells during nitrogen limitation, and 3) inactive cells formed by differentiation of the storage cells. Both the basic and the storage cells grow, whereas the inactive cells are formed only by differentiation. Citric acid is formed only in the storage cells. The growth of the basic cells is described by Monod kinetics with the nitrogen being the limiting substrate, and growth of the storage cells is described by Contois kinetics with sugar as the limiting substrate. Despite the morphologically structuring of the biomass the model has an empirical nature since it is difficult to interpret the biological nature of the individual morphological forms. In an empirical fashion the model simulates the citric acid production process quite well but it does not reveal much of the mechanisms behind the process.

Nestaas and Wang [71] describe a simple morphologically structured model, which was used to simulate the production of penicillin by *Penicillium chrysogenum*.

Table 5. The model of Nestaas and Wang [71]

Growth reactions			
$\alpha_1 S \rightarrow Z_H$	$r_1 = k_1 Z_A$	tip extension	(23)
$\alpha_2 S \rightarrow Z_A$	$r_2 = k_2 Z_A$	branching	(24)
Metamorphosis reaction			
$Z_H \rightarrow Z_I$	$u_1 = k_3 Z_H$	differentiation	(25)

Three morphological forms are considered: 1) Apical compartment (Z_A), 2) subapical compartment (Z_H), and 3) inactive cells (Z_I). The model is summarized in Table 5. Z_A and Z_H are synthesized directly from the substrate (23) and (24), and degeneration to Z_I is described by the metamorphosis reaction (25). The size of the apical compartment is assumed to be constant and tip extension therefore results in formation of Z_H, whereas Z_A is formed only by branching. There is no influence of the substrate in the model, but this may easily be introduced by specifying k_i, $i = 1, 2, 3$ as functions of S, e.g. with Monod expressions.

In the Nestaas and Wang model, it is assumed that branching occurs apically and r_2 is therefore taken to be a function of Z_A rather than of Z_H. As discussed in Sect. 4.3, apical branching is rarely observed but the model could easily be modified to describe branching in the subapical compartment by replacing r_2 in Eq. (24) with: $r_2 = k_2 Z_H$. The specific growth rate of the biomass is given by Eq. (26) and the mass balance for Z_A by Eq. (27).

$$\mu = (k_1 + k_2) Z_A \tag{26}$$

$$\frac{dZ_A}{dt} = r_2 - \mu Z_A = (k_2 - (k_1 + k_2) Z_A) Z_A \tag{27}$$

From Eq. (27) it can be seen that the steady state level of Z_A is given by $k_2/(k_1 + k_2)$. If k_i, $i = 1, 2, 3$ are specified as functions of S, the model may predict an increasing Z_A with increasing substrate concentration — just like the Megee et al. model. Product formation is included in the model by assuming that only the subapical cells can synthesize penicillin. The model corresponds quite well with experimental data, but a drawback is that an influence of the substrate is not included neither in r_1 nor in r_2, and it was therefore necessary to change the parameter values at a certain time when fed-batch operation was simulated. However, the model has a very simple structure and it is an attractive alternative to an unstructured model since the number of parameters is kept low.

5.2 Intracellularly Structured Models

In the morphologically structured models the kinetics is described only as a function of the environment and the fractional concentrations of the morphological forms. These models have much more predictive strength than unstructured models, but for a good description of rapid changes in the environmental conditions it may be necessary to structure the interior of each morphological form. This can be done by including key intracellular components in the model, e.g. the ribosome content as discussed in Sect. 3.3. The strength of intracellularly structured models is illustrated by the model of Imanaka et al. [75], which describes the production of α-galactosidase by a *Monasscus* mould. The model is able to describe multiple steady states in a chemostat and many other observations made during growth and product formation with the applied microorganism [75]. The model has recently been reviewed [5] and it is therefore not discussed further.

Combination of intracellular and morphological structure is illustrated in a model by Matsumura et al. [86] which is used to simulate cephalosporin C production by *Cephalosporium acremonium*. Three morphological forms are considered: 1) Hyphae, 2) swollen hyphae, and 3) arthrospores. Only the hyphae grow, whereas the swollen hyphae and the arthrospores are formed from respectively hyphae and swollen hyphae by metamorphosis reactions. Cephalosporin C is synthesized only by the swollen hyphae. Intracellular methionine is included in the model since this substrate component was found to have a profound influence on the product formation kinetics. The growth of the hyphae is described by Monod kinetics, whereas the rate of the metamorphosis reactions is described by simple functions of the substrate concentrations (both glucose and methionine) [5, 86]. The model is an illustration of how a morphologically structured model is combined with a mechanistic model for the product formation.

Prosser and Trinci [72] developed a computer model which includes most of the mechanisms discussed in Chapter 4. The model is formulated in terms of discrete events and furthermore, the model is constructed to describe the growth of one hyphal element only. In the model it is assumed that vesicles necessary for tip extension are synthesized throughout the apical compartment. The vesicles are transported to the apex where they fuse with the cell wall and hereby give rise to tip extension. Branching is included in the model by a mechanism where a new tip is formed when the level of vesicles somewhere in the hyphae attains a critical value. Formation of a septum is described by the introduction of the duplication cycle [50] (see Sect. 4.3). The computer model corresponds very well with experimental data for a single hyphal element, and it was valuable for examination of hypotheses concerning the growth of filamentous fungi. A drawback of the model is, however, that it is a verbal model only, and its application for simulation of submerged growth is therefore limited. Furthermore, an influence of the limiting substrate is not included.

Aynsley et al. [73] have tried to derive a quantitative model based on the concepts of the model of Prosser and Trinci. The vesicle production rate along the hypha is described by Monod kinetics, and a constant flow of vesicles along the hypha is assumed. In the model the hypha is regarded as a form of a self extending tubular reactor, and by assuming that the intracellular vesicle concentration is in a pseudo steady state at any position in the hypha Aynsley et al. derived an expression for the average concentration of vesicles (kg vesicles per kg biomass) at the hyphal tips in a submerged culture. The rate of absorption of the vesicles at the tips, equal to r_{tip}, is described by Monod kinetics. By using Eq. (9) an expression for the specific growth rate was derived by Aynsley et al.:

$$\mu = k_1 k_2 \frac{s}{K_1 K_2 + (K_2 + k_1 x_{hgu}) S} \tag{28}$$

The model is based on a number of assumptions which are doubtful, e.g. the flow of vesicles through the hypha is probably not constant and presence of pseudo steady state in the vesicle concentration at any position along the hypha is very dubious. The model reveals little of the growth mechanisms, and its predictive strength is uncertain. The expression for the specific growth rate in Eq. (28) can therefore be considered nothing more than an empirical expression with no biological background.

6 Submerged Growth of Filamentous Fungi

During submerged growth many strains of filamentous fungi may grow both as pellets and in a filamentous form, and one may therefore distinguish between morphology at two levels: microscopic morphology and macroscopic morphology. The microscopic morphology characterizes the hyphal elements, e.g. the total length and the number of tips, and the macroscopic morphology describes whether the hyphal elements are agglomerated into pellets or separated from each other. Pellets are characterized by tightly branched hyphae and a coupling between the micro- and macroscopic morphology therefore exists.

In a submerged culture the hyphal elements are exposed to shear forces which may result in fragmentation, i.e. break up of the hyphal elements. Fragmentation influences the observed microscopic morphology and it must therefore be considered when experimental data from submerged growth are interpreted. During growth of filamentous fungi in the form of pellets, diffusion of substrate components into the pellet may become the rate limiting process, and toxic metabolic products may accumulate inside the pellets. It is also necessary to consider this aspect when a process involving pellets is to be modelled.

A submerged culture of filamentous fungi shows non-Newtonian behavior, i.e. the apparent viscosity changes with the shear rate, and the apparent viscosity is normally quite high. Most media are characterized as pseudoplastic, i.e. the apparent viscosity decreases with the shear rate, and the most often applied rheological models are the Bingham and the Casson Eq. [74]. A high apparent viscosity has a significant influence on both the flow behaviour in the bioreactor and on the mass transfer, e.g. mass transfer of oxygen from the gas to the liquid phase [2]. It is therefore of interest to predict the apparent viscosity during a growth process. In the literature, a number of empirical rheological models [74] are described and they include both the total biomass concentration and microscopic morphological variables such as the hyphal growth unit.

6.1 Fragmentation of Hyphal Elements

Fragmentation is a prerequisite for obtaining true steady states in continuous cultures of filamentous fungi, since new hyphal elements are formed only due to break up. Fragmentation influences the observed microscopic morphology since two new tips are formed upon breakage of a hyphal element. The newly formed tips do not grow, and one should therefore ideally distinguish between tips formed by branching and upon fragmentation.

Fragmentation of a hyphal element takes place when the local shearing forces become larger than the tensile strength of the hypha. The rate of fragmentation is therefore a function of the number of times the hyphal element comes into the impeller zone where the local turbulent shearing forces are larger than the tensile strength of the hypha. Ideally all hyphal elements in a submerged culture would have the same size and morphology, but due to fragmentation and to the different behavior of the individual hyphal elements there will be a distribution of hyphal elements. According to Ramkrishna [75, 76] the population of hyphal elements is given by Eq. (29) where ε dm dn is the number of hyphal elements with a mass between m and m + dm and a tip number between n and n + dn. b is the rate of formation of new hyphal elements upon fragmentation (hyphal elements formed $(m^{-3} h^{-1})$ and D is the dilution rate in the bioreactor.

$$\frac{\partial \varepsilon(n, m)}{\partial t} + m \frac{\partial}{\partial n} (\varphi \varepsilon(n, m)) + n \frac{\partial}{\partial m} (r_{tip} \varepsilon(n, m))$$

$$= b(m, n, \ldots) - D\varepsilon(n, m) \tag{29}$$

The rate of fragmentation b in Eq. (29) is a complex function of both the distribution function of hyphal elements, $\varepsilon(n, m)$, and of the environmental conditions, e.g. the shearing forces acting on the hyphal elements. The population balance Eq. (29) is rarely applied due to the mathematical complexity and since correct specification of b is difficult. However, from measurements of the distribution function $\varepsilon(n, m)$ [65] one may examine different functions for b and hereby obtain information of the fragmentation process. In such a study one may benefit from models describing crystallization processes [76].

A simpler model based on average values would be valuable. The population of hyphal elements is completely homogeneous if it is assumed that the fragmentation process is synchronized and that the break up always results in formation of two hyphal elements of equal size. These assumptions are indeed very crude, but they may be accepted as a first attempt. For the total number of hyphal elements one finds:

$$\frac{d\varepsilon}{dt} = b - D\varepsilon \tag{30}$$

Since fragmentation results in the formation of two new tips a balance for the total number of tips in the culture is given by:

$$\frac{dn\,\varepsilon}{dt} = \varepsilon\,\frac{dn}{dt} + n\,\frac{d\varepsilon}{dt} = \varphi m\varepsilon + 2b - Dn\varepsilon \tag{31}$$

and by inserting Eq. (30) a balance for the number of tips on each hyphal element is obtained:

$$\frac{dn}{dt} = \varphi m + (2 - n)\,\frac{b}{\varepsilon} = \varphi x_{hgu} n + (2 - n)\,\frac{b}{\varepsilon} \tag{32}$$

At steady state one finds:

$$n = \frac{2b}{b - \varphi x_{hgu}\varepsilon}; \qquad \varepsilon = \frac{b}{D} \tag{33}$$

It is often suggested that fragmentation in a submerged culture is determined mainly by the shear forces imparted by the impeller, but other factors, e.g. the substrate availability may be of importance. It is likely that older hyphae fragment more easily than new cells which may have a very rigid cell wall, and the fragmentation may therefore also be coupled to the differentiation processes. Addition of reagents which induce differentiation have been found to result in an increased fragmentation in a submerged culture [77]. In a glucose-limited chemostat Metz et al. [65] found that both the total hyphal length and the average number of tips increased with the specific growth rate. The hyphal diameter was approximately constant and the total hyphal length is therefore proportional to m. The total biomass concentration was almost constant, and since m increases with dilution rate D, it is obvious that $\varepsilon = x/m$ must decrease with increasing D. The data indicate that m is proportional to D^a, where $a > 1$, and by using Eq. (33) it is found that b is proportional to D^{1-a}, i.e. b is a decreasing function of D. The experimental data for n vs dilution rate reveal a similar trend — again by using Eq. (33). It is therefore concluded that the breakage function is a decreasing function of D — or more likely: a decreasing function of the glucose concentration, which may be explained by the formation of a more rigid cell wall with higher tensile strength at high glucose concentrations. Variation in the breakage function with the glucose concentration may explain why the hyphal diameter is constant in the experiments reported by Metz et al. [65] whereas the general observation is that the diameter increases with D [59] (see Sect. 4.3). The influence of other environmental factors on the breakage function, e.g. pH and dissolved oxygen, have also been examined, but these factors do not seem to have a pronounced effect on the microscopic morphology [78].

The breakage function is obviously a function of the shearing forces and Suijdam and Metz [78] have derived a model for b, which is based on engineering theories for dispersion of physical systems. In the model, b is a function of the impeller dimensions, the agitation rate, and morphological variables like the hyphal diameter, the total length of the hyphal element, and the tensile strength of the hypha. The model corresponds qualitatively with experimental data for the microscopic morphology in a chemostat with varying energy input, i.e. agitation rate, and constant specific growth rate. The experiments show that the total hyphal length, hyphal growth unit length, and hyphal diameter decrease when the stirring speed is increased. A similar observation is made elsewhere [79]. According to Eq. (10) the hyphal growth unit mass therefore decreases with the energy input. This contradicts the hypothesis of Sect. 4.3 that x_{hgu} is constant for a given strain. With a high degree of fragmentation in the submerged culture a large number of inactive tips (n_i) are formed, and these cannot be distinguished from the actively growing tips (n_a) with the applied experimental procedure. The larger number of inactive tips probably explains the decrease in l_{hgu} for increasing stirring speed. If the mass of hypha supporting the growth of one active tip is constant, i.e. $x_{hgu,a} = m/n_a$, the number of actively growing tips per hyphal element must decrease when the stirring speed is increased. When a hyphal element fragments in the apical compartment all the cytoplasmic material streams out to the medium, and an actively growing tip is converted to an inactive tip. A decrease in the number of actively growing tips with increasing degree of fragmentation therefore seems reasonable.

6.2 Macroscopic Morphology: Mycelial Pellets

Depending on the operating conditions, some filamentous fungi may form pellets when they are grown submerged. Pellet formation is determined by many different factors [80], e.g. the carbon-nitrogen ratio in the substrate and the spore concentration in the inoculum. Changes in the macroscopic morphology may occur in both directions, i.e. pellets may be formed and disappear again [81]. The growth kinetics of pellets is often described by the empirical cube-root growth law [82].

A pellet suspension has a lower apparent viscosity than a culture in the filamentous form, and problems with mixing and gas-liquid mass transfer are therefore reduced. With the application of pellets one does, however, face another mass transfer problem: Transport of substrates into the growing cells inside the pellet. The transport of species in pellets is normally by molecular diffusion. Assuming the pellets are spherical and the pellets have an uniform density the steady state mass balance Eq. (34) may be derived for the concentration profile inside the pellets. D_s is the diffusion coefficient, R is the pellet radius, and r_s is the specific substrate uptake rate. $x(\zeta, t)$ is the biomass concentration at the position ζ in the pellet.

$$\frac{D_s}{R^2}\left(\frac{\partial^2 s(\zeta, t)}{\partial \zeta^2} + \frac{2}{\zeta}\frac{\partial s(\zeta, t)}{\partial \zeta}\right) = \varrho_s x(\zeta, t) \tag{34}$$

The diffusion coefficient D_s is normally a function of the pellet density, i.e. the biomass concentration profile in the pellet. Equation (34) is based on an assumption of the transport being completely diffusive, which is the normal situation. For modelling a culture containing pellets of varying size and properties it is necessary to combine Eq. (34) with a description of the pellet population. Normally one will assume that all the pellets have an equal size whereby their size may be described by Eq. (35) where μ is the specific growth rate of the whole biomass. By assuming a pseudo steady state in the concentration profile of the limiting substrate, Eqs. (34) and (35) may, together with a model for μ and r_s, be used to simulate the growth of mycelial pellet cultures. This approach is illustrated in an unstructured model for mycelial pellets by Suijdam et al. [83].

$$\frac{dR}{dt} = \frac{1}{3}\mu R \tag{35}$$

To model pellet suspensions it is necessary to consider pellet break up, since this has a significant effect on the pellet diameter, i.e. an additional term is needed in Eq. (35). Fragmentation of pellets involves two shear dependent processes: chipping of pieces of hyphae from the outside of the pellet and total rupture of the pellet [84]. The first process results in a decreasing pellet diameter, and it has been found to be a function of the energy input and the pellet diameter. When non-growing pellets are placed in a stirred vessel the diameter decreases exponentially to a constant level which depends on the stirring rate in the vessel [85].

Acknowledgements: I am very grateful to Professor J. Villadsen, who made several useful suggestions regarding this article. I would also like to thank L. H. Christensen and L. B. Jørgensen for their encouraging discussions on the growth of filamentous fungi.

7 References

1. Nüesch J, Heim J, Treichler H-J (1987) Ann Rev Microbiol 41: 51
2. Schügerl K (1981) Adv Biochem Eng 19: 71
3. Moser A (1988) Bioprocess technology, Springer, Vienna New York
4. Schügerl K (1991) Bioreaction engineering. Characteristic features of bioreactors, vol 2, John Wiley, Chichester
5. Nielsen J, Villadsen J (1992) Chem Eng Sci, in press
6. Fredrickson AG (1976) Biotechnol Bioeng. 18: 1481
7. Roels JA (1982) Energetics and kinetics in biotechnology, Elsevier, Amsterdam
8. Nielsen J, Nikolajsen K, Villadsen J (1991) Biotechnol Bioeng 38: 1
9. Tsuchiya HM, Fredrickson AG, Aris R (1966) Adv Chem Eng 6: 125
10. Fredrickson AG, Ramkrishna D, Tsuchiya HM (1967) Math Biosci 1: 327
11. Måløe O, Kjeldgaard No (1966) Control of macromolecular synthesis, WA Benjamin, New York
12. Cooper S (1991) Bacterial Growth and Division, Academic, San Diego
13. Blumenthal HJ (1965) Glycolysis. In: Ainsworth GC, Sussman AS (eds)The fungi, vol 1, Academic, New York, p 229

222   J. Nielsen

14. Berry DR (1975) The environmental control of the physiology of filamentous fungi. In: Smith JE, Berry DR (eds) The Filamentous fungi, vol 1, Edward Arnold, London, p 16
15. Shu P, Funk A, Neish AC (1954) Can J Biochem Physiol 32: 68
16. Carter BLA, Bull AT (1969) Biotechnol Bioeng 11: 785
17. Lewis KF, Blumenthal HJ, Wenner CE, Weinhouse S (1954) Federation Proc. 13: 252
18. Heath EC, Koffler H (1956) J Bacteriol 71: 174
19. Wang CH, Stern I, Gilmour CM, Klungsoyr S, Reed DJ, Bialy JJ, Christensen BE, Cheldelin VH (1958) J Bacteriol 76: 207
20. Reed DJ, Wang CH (1959) Can J Microbiol 5: 59
21. Bull AT, Trinci APJ (1977) The Physiology and Metabolic Control of Fungal Growth. In: Rose AH, Tempest DW (ed) Adv Microbial Physiol 15: 2
22. Alexander MA, Jeffries TW (1990) Enzyme Microb Technol 12: 2
23. Niederpruem DJ (1965) Tricarboxylic acid cycle. In: Ainsworth GC, Sussman AS (eds) The fungi, vol 1, Academic, New York, p 269
24. Lindenmayer A (1965) Terminal oxidation and electron transport. In: Ainsworth GC, Sussman AS (eds) The fungi, vol 1, Academic, New York, p 301
25. Carter BLA, Bull AT, Pirt SJ, Rowley BI (1971) J Bacteriol 108: 309
26. Pirt SJ, Callow DS (1960) J Appl Bacteriol 23: 87
27. Pirt SJ, Righelato (1967) Appl Microbiol 15: 1284
28. Ryu DDY, Hospodka J (1980) Biotechnol. Bioeng. 22: 289
29. Mason HRS, Righelato RC (1976) J Appl Chem Biotechnol 26: 145
30. Sonnleitner B, Käppeli O (1986) Biotechnol Bioeng 28: 927
31. Hackette SL, Skye GE, Burton C, Segel IH (1970) J Biol Chem 245: 4241
32. Pateman JA, Kinghorn JR (1976) Nitrogen Metabolism. In Smith JE, Berry DR (ed) The Filamentous Fungi, vol 2, Edward Arnold, London: 159
33. Roos W, Luckner M (1984) J Gen Microbiol 130: 1007
34. Hunter DR, Segel IH (1971) Arch Biochem Biophys 144: 168
35. Hunter DR, Segel IH (1973) J Bacteriol 113: 1184
36. Casselton PJ (1976) Anaplerotic pathways. In: Smith JE, Berry DR (ed) The Filamentous Fungi, vol 2, Edward Arnold, London: 121
37. Steinmeyer DE, Shuler ML (1989) Chem Eng Sci 44: 2017
38. Nielsen J, Nikolajsen K, Villadsen J (1991) Biotechnol Bioeng 38: 11
39. Alberghina L, Sturani E, Costantini MG, Martegani E, Zippel R (1979) In: Burnett JH, Trinci APJ (ed) Fungal walls and hyphal growth, Cambridge University Press, Cambridge: 295
40. Sturani E, Magnani F, Alberghina FAM (1973) Biochim Biophys Acta 319: 153
41. Bushell ME, Bull AT (1976) J Appl Chem Biotechnol 26: 339
42. Ingraham JL, Maaløe O, Neidhardt FC (1983) Growth of the bacterial cell, Sinnauer Associates, Sunderland
43. Righelato RC (1975) Growth kinetics of mycelial fungi. In: Smith JE, Berry DR (ed) The filamentous fungi, vol 1, Edward Arnold, London: 79
44. Righelato RC, Trinci APJ, Pirt SJ and Peat A (1968) J Gen Microbiol 50: 399
45. Smith JE (1975) The structure and development of filamentous fungi. In: Smith JE, Berry DR (ed) The filamentous fungi, vol 1, Edward Arnold, London: 1
46. Bartnicki-Garcia S (1990) Role of vesicles in apical growth and a new mathematical model of hyphal morphogenesis. In: Heath IB (ed) Tip growth in plant and fungal cells, Academic, San Diego: 211
47. Smith JE (1978) Asexual sporulation in filamentous fungi. In: Smith JE, Berry DR (eds) The filamentous fungi, vol 3, Edward Arnold, London: 214
48. Steele GC, Trinci APJ (1975) J Gen Microbiol 91: 362
49. Fiddy C, Trinci APJ (1976) J Gen Microbiol 97: 169
50. Trinci APJ (1978) The duplication cycle and vegetative development in moulds. In: Smith JE, Berry DR (ed) The filamentous fungi, vol 3, Edward Arnold, London: 132
51. Trinci APJ, Collinge AJ (1975) J Gen Microbiol 91: 355
52. Bartnicki-Garcia S (1973) Symp Soc Gen Microbiol 23: 245

53. Prosser JI (1979) Mathematical modelling of mycelial growth. In: Burnett JH, Trinci APJ (ed) Fungal walls and hyphal growth, Cambridge University Press, Cambridge: 359
54. Grove SN (1978) The cytology of hyphal tip growth. In Smith JE, Berry DR (ed) The filamentous fungi, vol 3, Edward Arnold, London: 28
55. Trinci APJ (1971) J Gen Microbiol 67: 325
56. Trinci APJ (1979) The duplication cycle and branching in fungi. In Burnett JH, Trinci APJ (ed) Fungal walls and hyphal growth, Cambridge University Press, Cambridge: 319
57. Trinci APJ (1984) Regulation of hyphal branching and hyphal orientation. In Jennings DH, Rayner ADM (ed) The ecology and physiology of the fungal mycelium, Cambridge University Press, Cambridge: 23
58. Robson GD, Wiebe MG, Trinci APJ (1991) J Gen Microbiol 137 963
59. Robinson PM, Smith JM (1979) Trans Br Mycol Soc 72: 39
60. Caldwell IY, Trinci APJ (1973) Arch Mikrobiol 88: 1
61. Katz D, Goldstein D, Rosenberger RF (1972) J Bacteriol 109: 1097
62. Morrison KB, Righelato RC (1974) J Gen Microbiol 81: 517
63. Kretschmer S (1985) J Basic Microbiol 25: 569
64. Riesenberg D, Bergter F (1979) Z Allgemeine Mikrobiol 19: 415
65. Metz B, Bruijn EW de, Suijdam JC van (1981) Biotechnol Bioeng 23: 149
66. Wiebe MG, Trinci APJ (1991) Biotechnol Bioeng 38: 75
67. Bajpai RK, Reuss M (1980) J Chem Tech Biotechnol 30: 332
68. Bajpai RK, Reuss M (1981) Biotechnol Bioeng 23: 717
69. Megee RD, Kinoshita S, Fredrickson AG, Tsuchiya HM (1970) Biotechnol Bioeng 12: 771
70. Kristiansen B, Sinclair CG (1979) Biotechnol Bioeng 21: 297
71. Nestaas E, Wang DIC (1983) Biotechnol Bioeng 25: 781
72. Prosser JI, Trinci APJ (1979) J Gen Microbiol 111: 153
73. Aynsley M, Ward AC, Wright AR (1990) Biotechnol Bioeng 35: 820
74. Metz B, Kossen NWF, Suijdam JC van (1979) Adv Biochem Eng 11: 103
75. Ramkrishna D (1985) Rev Chem Eng 3: 49
76. Ramkrishna D (1979) Adv Biochem Eng 11: 1
77. Drew SW, Winstanley DJ, Demain AL (1976) Appl Environ Microbiol 31: 143
78. Suijdam JC van, Metz B (1981) Biotechnol Bioeng 23: 111
79. Miles EA, Trinci APJ (1983) Trans Br Mycol Soc 81: 193
80. Metz B, Kossen NWF (1977) Biotechnol Bioeng 19: 781
81. Pirt SJ, Callow DS (1959) Nature 4683: 307
82. Trinci APJ (1970) Arch Mikrobiol 73: 353
83. Suijdam JC van, Hols H, Kossen NWF (1982) Biotechnol Bioeng. 24: 177
84. Tagushi H (1971) Adv Biochem Eng. 1: 1
85. Suijdam JC van, Metz B (1981) J Ferment Technol 59: 329
86. Matsumura M, Imanaka T, Yoshida T, Taguchi H (1981) J Ferment Technol 59: 115

An Overview of the Biotechnology Research Activities in the European Community

I. Economidis
Commission of the European Communities, Directorate General for Science, Research and Development, Directorate Biology, Division Biotechnology, Rue de la Loi 200, B-1049 Brussels

The research in Biotechnology is an important element for the competitive environment of European bioindustrial activities. In this article the evolution of the Biotechnology programmes in the European Community is presented.

1 Introduction

A recent communication [1] of the Commission of the European Communities to Parliament and to Council starts by saying that

> "Biotechnology is a key technology for the future competitive development of the Community and it will determine the extent to which a large number of industrial activities located within the Community will be leaders in the development of innovatory products and processes."

The Commission recognises that strengthening the scientific and technological base of the industry is essential for the Community's industries to become more competitive at international level. The Community's principal role is to furnish the necessary dynamism and coherence, to contribute to the definition of joint projects, to the coordination of the various interests involved, to the exchange and diffusion of results and to the harmonisation of actions lying within its competence. The Single European Act [2], which brought research and tech-

Advances in Biochemical Engineering
Biotechnology, Vol. 46
Managing Editor: A. Fiechter
© Springer-Verlag Berlin Heidelberg 1992

Table 1. The European Community's stategy for research and technological development

---

Article 130 F

1. The Community's aim shall be to strengthen the scientific and technological basis of European industry and to encourage it to become more competitive at international level.

2. In order to achieve this, it shall encourage undertakings including small and medium-sized undertakings, research centres and universities in their research and technological development activities; it shall support their efforts to cooperate with one another, aiming, in particular, at enabling undertakings to exploit the Community's internal market potential to the full, in particular through the opening up of national public contracts, the definition of common standards and the removal of legal and fiscal barriers to that cooperation.

3. In the achievement of these aims, particular account shall be taken of the connection between the common research and technological development effort, the establishment of the internal market and the implementation of common policies, particularly as regards competition and trade.

Single European Act

---

nological development for the first time explicitly into the EEC Treaty, has provided new impetus towards an overall strategy for research and competitiveness in the bio-industries (Table 1).

2 The Biotechnology Research Programmes in the European Community (1982 – 1993)

2.1 The Pioneer Programme: BEP [3]

The Biomolecular Engineering Programme (BEP) was the first Community initiative in the field of Biotechnology. Its objectives were to contribute, through transnational efforts in research and training, to the removal of bottlenecks which inhibit the application of molecular and cellular biology to agriculture and to the agro-food industries. With a budget of 15 Mio ECU, BEP supported 91 training contracts and 103 cost-shared research contracts with public and private laboratories in the Community. In the area of genetic engineering, several significant contributions, such as a gene cloning in *Streptococcus* bacteria exploited by the cheese industries, the characterization and isolation of more than 20 plant genes of great importance for agriculture or genetic transformation in plant species belonging to the monocots, resulted from the programme, 53 cooperative agreements were established between laboratories in different member states which involve the exchange of equipment and staff and the execution of joint experiments. Permanent transnational working parties have been set up for each research sector in the programme.

2.2 BAP: The Second R & D Programme in Support of Biotechnology [4]

The need to reinforce the Community's R & D effort [2] was recognized in the 1985 – 1989 "Biotechnology Action Programme (BAP)" overlapping BEP and with resources of 55 Mio ECU which were further increased by 20 Mio ECU under the frame of a programme revision (1987). The programme, inherently non-competitive, was oriented towards medium- and long-term objectives (promotion of research, training and concerted action at the Community level) deemed necessary for ensuring strategic strength of European industry and agriculture. With regard to research and training, BAP focused upon two essential tasks, namely: 1) the establishment of a supportive intrastructure for biotechnology research and development in Europe; 2) the elimination of bottlenecks which prevented the exploitation by industry and agriculture of the materials and methods originating from modern biology.

The amplification of BAP (the programme revision) not only enabled Spanish and Portuguese laboratories to join the activity in mid-programme, but allowed two significant developments:

— amplification of safety assessment work, to involve more than 58 laboratories throughout the Community: to cope with rising political concerns about conjectural risks (especially in the agricultural and environmental use of modified organisms), and to provide a scientific basis for possible regulation;
— reinforcement of bio-informatics activities in Europe (e.g. the DNA sequence library at the European Molecular Biology Laboratory, Heidelberg), in response (along with the AIM programme, of Advanced Informatics in Medicine) to recommendations arising from the BICEPS initiative of exploratory studies and workshops (Bio-informatics: Collaborative European Programmes and Strategy).

From a total of 1357 applications submitted in response to the call for proposals launched in 1985, 262 2–4 year shared-cost research contracts, divided into 93 projects, were negotiated by the Commission Services. The revision of BAP made additional funds available to support 116 new laboratories during 1989–90. These laboratories either initiated new joint projects or were integrated into the pre-existing 93 original BAP projects. The BAP projects (292 laboratories) have been classified into 9 sectors of priority research, corresponding to a functional rearrangement of the 20 specific research lines initially composing the scope of the programme (Table 2).

Table 2. BAP

Scope of the programme	Major research sectors
Contextual measures	
Bio-informatics	bio-informatics
Data capture technologies	
Updating and design of data banks related to biotic materials	
Modelling techniques and algorithms	
Advanced computer software, including expert systems	

Table 2. (continued)

Scope of the programme	Major research sectors
Collections of biotic materials	culture collections
Upgrading of existing collections of importance as supporting resources for biotechnology R & D, and creation of new collections, required and made possible by the advances of science	
Development and improvement of technical methods of storage and resuscitation as well as improved methods of identification description and classification	
Basic biotechnology	
Enzyme engineering	
Development of bioreactors (multi-enzymatic, multi-phasic, co-factor requiring or utilising viscous media)	bioreactors
Stability of enzymes during industrial exploitation Protein design including new concepts enzyme catalysis, structural and functional predictions, chemical and genetic modifications, constructions of artificial enzymes	protein design/ macromolecular modelling
Genetic engineering	
Microorganisms: gene characterisation and gene transfer for potential applications by industries	biotechnology of industrial microorganisms
Plants: analyses of the structure and regulation of plantgenomes	biotechnology of plants and assoc. microorganisms
Plants: transfer and cloning of genetic material in microorganisms and plants: association (in particular symbiotic relations) between crop plants and micro-organisms	
Plants: early detection of genetic or pathological modifications in cultivated plants	
Animals (livestock, including fish): cloning of substances important for animal husbandry, cloning vectors for animal cells, cloning of genetic material in animal cells	biotechnology of animals
Technology of cells and tissues cultured in vitro	
Physiological and genetic factors governing yield and stability during continuous cultivation of microbial species important to industry	biotechnology of industrial microorganisms (cont)
Control of the differentiation of plant cells and of their regeneration in entire plants	biotechnology of plants & assoc. microorganisms (cont)
Novel methodologies of animal cell cultures	biotechnology of animals (cont)
Assessment of risks	risk assessment
Development of new methods for detecting contamination and for the assessment of possible risks associated with applications in industry and agriculture of biomolecular engineering. In particular, development of methods for the assessment of risks resulting from the release of genetically engineered organisms	
Testing methods	in vitro testing methods
The development of in vitro screening methods for the evaluation of the toxicological effects and the biological activity of molecules	

2.3 European Laboratories Without Walls (ELWW) [5]

The collaborative links which the Commission attempted to promote and stimulate throughout the duration of the programme materialized in the creation of 35 European Laboratories Without Walls (ELWW), which represented the ultimate state of cooperation and integration of research efforts in Community biotechnology.

An ELWW is an open-ended, transnational association of cooperating European groups (laboratories to universities, institutes or industries) with a common commitment to target-oriented multi-disciplinary research. The basic concept is to involve first-class research scientists in their disciplines and to encourage the integration of their work through contractual arrangements with the European Commission aiming towards the solution to specific problems or towards the acquisition of new knowledge or understanding of basic processes important for the exploitation of biological material. Within the ELWW there is a premium on information exchange and technological transfer. The information generated by research is circulated freely, creating a framework for the continuous exchange of data, materials and staff, especially between academic and industrial scientists. Particular importance is attached to the exchange of scientific staff to be trained in specific sophisticated techniques. Regular joint meetings are held in order to evaluate and discuss the results obtained, and to plan future experiments.

By receiving an input from various European countries, each ELWW benefits from a variety of research approaches and expertise across a wide range of disciplines. This gives an added benefit to European biotechnology since ELWWs can provide combinations of skills rarely available within the confines of an individual university, institute or company; and often not available in a single country.

A typical ELWW might consist of six research groups from universities or public research institutes, together with three industrial companies. A variety of experimental activities and scientific exchanges taking place all the time, but full joint meetings take place only every six months in a different host laboratory. Researchers from some of the laboratories will normally spend a number of weeks in another laboratory to learn a specific technique and perform joint experiments. Most of the groups will have exchanged biological materials, and several joint papers will have been published in addition to those of individual laboratories.

2.4 BRIDGE (1990–1993): The Current Programme [6, 7]

BRIDGE (Biotechnology, Research for Innovation, Development and Growth in Europe) is consolidating the developments achieved in BAP — industrial involvement, contextual measures, concerted action — it also displays novel features, of which three can be briefly mentioned:

- the programme is open to EFTA countries;
- in line with advice from industry, and the positive Evaluation Panel report on BEP and BAP, the programme is implemented by means of "N-projects",

network-based activities of typically more academic character and modest financial scale, and "T-projects", more targeted in character, multimillion in financial cost;
— building on a small start in BAP, genome analysis work will be pursued on both yeast and the plant *Arabidopsis* — both ideal models for study — in conjunction with world-wide collaboration.

The priorities of BRIDGE (Table 3) are based on:

— the necessity of contributing to the modernisation of the infrastructures of data banks and culture collections in the Community;
— the importance for agriculture and industry, underlined by all consultants, of fundamental research in protein design, gene analysis, and in the molecular and cellular biology of species exploited in agriculture and industry;
— need to develop a Community strength in the area of pre-normative research (research where norms do not, as yet exist) (development of in vitro tests for the evaluation of the properties of new molecules and assessment of risks possibly associated to the release in the environment of genetically manipulated organisms — GMO's).

Table 3. BRIDGE

BRIDGE	Number of laboratories selected
Information Infrastructure (Processing and analysis of biological data)	31
Enabling Technologies (Protein design; molecular modelling; bio-transformation; genome sequencing)	158
Cellular Biology (industrial microorganisms, plants and associatged microorganisms; animal cells)	223
Pre-Normative Research (in vitro evaluation of the activity of molecules; safety assessment of GMO's)	107
Total	519

2.5 The T-Projects of BRIDGE: A New Tool for Technology Transfer in the Community [8]

The concept of T-projects derived from a suggestion by a group of independent experts who foresaw in Community programmes for Biotechnology the opportunity to implement large-scale projects directed towards the solution of major trans-European problems. The group suggested a number of targeted objectives which were all taken up by BRIDGE and are to be maintained in the new Biotechnology programme which the Commission services hope to launch in 1992.

In Bridge, T-projects are now under way for the removal, through a significant investment of skills and resources, of important bottlenecks resulting from structural and scale constraints. The precise definition of each T-project was established after discussion with experts nominated by the Member States and through consultation of the committees who advise the Commission during the preparation of new specific programmes (Table 4). Each T-project is implemented

Table 4. The T-projects of BRIDGE

1	Sequencing of yeast genome:	Sequencing of chromosome II (840 kb) and chromosome XI (680 kb) in *S. cerevisiae*, a species chosen as test material for industrial microorganisms and as a model for eukaryotes.
2	Molecular identification of new plant genes:	Characterisation in *Arabidopsis thaliana*, for subsequent search and identification in crop plants, of genes not accessible through classical methods which are involved in fundamental processes of crop physiology.
3	Biotechnology of lactic acid bacteria:	Genetics, molecular biology, physiology and biochemistry of lactic acid bacteria with special reference to the understanding and control of industrial properties and to the use of modern techniques for the screening of new strains.
4	Industrial lipases	Structural and functional characterisation of 10–15 lipases with the aim of obtaining the basic understanding of their functions. This new knowledge about a sufficient number of these enzymes will make it possible to understand why they are lipases and how they function as such.
5	Regulation of plant cell generation:	Analysis of mechanisms by which a diversity of signals controls growth and cell differentiation in higher plants; development of tools, at cellular and molecular levels, for the study of morphogenesis
6	High resolution automated microbial identification:	Development of technologies and instrumentation which increase the rapidity, sensitivity and specificity of microbial identification, decrease its labour intensiveness and increase its cost effectiveness. The results of the research should be essential for the improvement of experimental approaches in risk assessment particularly in the monitoring of released genetically modified organisms and for assessing their possible impact on indigenous species in the target ecosystems.
7	Animal cell technology	Development of host-vector systems for animal cell biotechnology adapted to the requirements for specificity, efficient expression, stability and large scale exploitation.

by a team of research laboratories selected after a specific call for transnational proposal. The number of laboratories in the team depends upon the nature of the project, within a range of 10 to 50. The contribution of the Community covers about half of the total cost of the project and varies between 1 and 2 Mio ECU/year/project.

3 "Biotechnology", A New Programme Now in Preparation

In May 1990 the European Commission proposed to Council a specific research and technological development programme in the field of Biotechnology [9] which is intended to amplify and complement some of the research efforts currently implemented in BRIDGE.

The scientific and technical objectives and content of this proposal have taken into account

— the approach embodied in the third Framework Programme in terms of the scientific and technical goals and the underlying aims which it pursues,
— the need for pre-normative research with emphasis on the safety assessment of new techniques and novel products,
— the ethical and social implications of Biotechnology,
— the importance of information technology for collecting, pooling, analyzing, distributing or simulating data.

Research is to be carried out at the level of molecules, cells, organisms and populations:
Molecular Approaches
Protein Structure and Function; structure and function of genes; expression of genes.
Cellular and Organism Approaches
Cellular regeneration; reproduction and development of living organisms; metabolism of animals; plants and microbes; essential physiological traits, communication systems within living matter.
Ecology and Population Biology
Ecological implications of Biotechnology; conservation of genetic resources.

4 The Research in Biotechnology as an Important Element for the Competitive Environment of the Bio-Industrial Activities

I started this article by mentioning the opening statement of the Commission's communication for promoting the competitive environment for the industrial activities based on Biotechnology within the Community.

I would like to finish by making reference to the recommendations made by the Commission of the European Communities for enhancing competitiveness and public acceptability.

1. the Community's contribution to research and development in the area of biotechnology should be reinforced. This will be undertaken in the review of the R & D framework programme;
2. the Community will, through its research programmes, information market policy, and international collaboration, contribute to the development of a biotechnology information infrastructure within the Community and world-wide (including data banks, software, and electronic networks and services);
3. in order that work in the field of standards may fully complement the Community's legislative work, a clear and precise mandate shall be prepared by the Commission's services, in consultation with CEN;
4. Community legislation currently under discussion in the area of intellectual property should be adopted, and Community legislation already adopted should be transposed into the legislation of the Member States, as a matter of urgency in order that the Community will have a coordinated approach which will strengthen its position in international negotiations.
5. statistics specific to biotechnology should be compiled in order that statistical monitoring of developments in the industrial application of biotechnology may take place;
6. bilateral and multilateral international contacts must be further strengthened. In addition to this the Community should pursue, within the context of international bilateral working groups, GATT, the OECD, EFTA and, where appropriate, other international bodies, the establishment of environmental and health objectives and should ensure that these are integrated into economic and other policy decisions.
7. to enable ethical issues to be clearly identified and discussed, the appropriate advisory structure at Community level should be established;
8. the Commission will regularly evaluate the progress and competitiveness of the biotechnology industries in Europe in order to make sure that the agreed framework remains appropriate. Success in this regard will, essentially, depend on the strategies adopted by the industries concerned.

In all eight proposed activities we see that research in biotechnology plays a key role. Research in Biotechnology is also the basic element for developing and implementing the Community's regulatory framework ensuring safety for man and the environment. The relevant legislation will require adaptation to technical progress and the progress of scientific knowledge, in order to deal with advances in biotechnology.

Acknowledgment: Prof. K. Schügerl participated in BEP with a project on the development of liquid membrane reactors for the production of 1-amino acids.

In BAP, Prof. Schügerl participated in two projects; one on the bioconversion of hydrophilic and hydrophobic compounds by enzyme systems and the other on the biochemical-biophysical knowledge, control and optimization of animal cell cultures in reactors.

His contribution was very much appreciated by the services of the Commission.

5 References

1. Promoting the Competitive Environment for the Industrial Activities based on Bio-technology within the Community, EEC, SEC (91) 629. final. Brussels, 1991
2. Treaties establishing the European Communities. ECSC-EEC-EAEC, Brussels–Luxembourg, 1987
3. Magnien E (ed) (1986) Biomolecular Engineering in the European Community. Achievements of the Research Programme (1982–1986) – Final Report. Martinus Nijhoff Pub Dordecht (NL) 1986
4. Vassarotti A, Magnien E (eds) (1990) Biotechnology R & D in the EC. Elsevier, Amsterdam
5. Magnien E, Aguilar A, Wragg P, de Nettancourt D (1989) A new tool for Biotechnology R & D in the Community. Biofutur Nov 1989
6. BRIDGE, Biotechnology Research for Innovation Development and Growth in Europe (1990–94) COM (88) 806 final – SYN 182, Brussels, 1989
7. de Nettancourt D (1991). BRIDGE: Launching of the research activities. Chimica Oggi April 1991
8. de Nettancourt D (1991) The T-projects of BRIDGE, a new tool for technology transfer in the Community, Agro-Industry hi-tech. 1991
9. Proposal for a Council Decision adopting a specific research and technological development programme in the field of Biotechnology (1990–94). COM (90) 160 final – SYN 265, Brussels, 1990

Author Index Volumes 1—46

Subject Index